NMR IN BIOLOGY

NMR IN BIOLOGY

Edited by

R. A. DWEK

Department of Biochemistry
University of Oxford
Oxford, England

I. D. CAMPBELL

Department of Biochemistry
University of Oxford
Oxford, England

R. E. RICHARDS

Merton College
Oxford, England

R. J. P. WILLIAMS

Department of Inorganic Chemistry
University of Oxford
Oxford, England

1977

ACADEMIC PRESS
London · New York · San Francisco

A Subsidiary of Harcourt Brace Jovanovich, Publishers

ACADEMIC PRESS INC. (LONDON) LTD.
24/28 Oval Road,
London NW1

United States Edition published by
ACADEMIC PRESS INC.
111 Fifth Avenue
New York, New York 10003

Library of Congress Catalog Card Number: 77-80291
ISBN: 0-12-225850-9

Printed in Great Britain by
The Scolar Press Ltd.
Ilkley, Yorkshire

PREFACE

This book is a record of the proceedings of the British
Biophysical Society's Spring Meeting on *NMR in Biology*
which was held in Oxford between 27 and 30 March 1977. The
aim of the meeting was to show how NMR spectroscopy could
be used to investigate biological problems at all levels
of organization — literally from molecules to men. The
study, in solution, of small molecules was largely omitted
since the use and value of NMR in organic and inorganic
chemistry is well recognized. (Abstracts 20-23 illustrate
how information can be obtained about the conformation of
relatively small molecules in solution.)

The first group of main papers in this book is devoted
to conformational features of proteins. In the last twenty
years this field has been dominated by X-ray diffraction
studies of single *crystals* and it is an important aim of
NMR studies that they should generate comparable information for proteins in *solution*. Bode *et al.* set the scene
by describing in detail the strengths and weaknesses of
the solid state approach. There is no doubt that diffraction techniques give the best over-view of a whole protein
and the most accurate interatomic coordinates, but it is
very largely restricted to a *static* picture of a protein in
a crystal, which is a non-biological salt medium and also
the *dynamic* properties of the protein are poorly described.
In the next three papers it is shown how NMR measurements
of interatomic distances in local regions of proteins indicate that the conformation in solution is closely related
to that in the crystal. The NMR method then adds to this
overall static picture the *conformational changes* which a

protein experiences from temperature dependent *fluctuations* or from changes in solution conditions, e.g., pH or added salt. The NMR method is seen at its best in comparative studies (see, for example, abstracts 3, 4, 7, 9 and 17). We note that non globular proteins (e.g. collagen) have not yet been very amenable to studies by X-ray techniques but they can be examined by NMR (Abstracts 10–12).

The next group of main papers illustrates the importance of dynamic as opposed to the static picture of molecules, especially for enzymes, which go through a series of conformational transitions on reaction and these can be followed by NMR. There is little doubt that this sort of dynamic information will lead to a new understanding of enzyme catalysis (see also abstracts 8, 18, 19).

There follows a series of papers on the combination of proteins with small molecules. The binding site of an antibody is defined using a variety of magnetic resonance methods while another paper attacks the ambitious problem of how a protein binds to DNA (see also abstracts 3, 6, 24, 34). Larger protein systems from muscle are considered in the paper by Edwards and Sykes, while other large systems and protein/protein interactions are discussed in abstracts 1–3, 5–11.

In the subsequent paper Robillard describes some of the solution studies on t-RNA while other studies of nucleic acids can be found in abstracts 35–38.

Increasing the scale of study from the level of solutions of large molecules leads in three different directions — (1) to organised single phases (membranes) (McLaughlin *et al.*, Abstracts 13, 39); (2) to biological space divided into different phases, e.g. simple cells and organelles (Daniels *et al.*), and (3) to whole organs (Gadian *et al.*, Dawson *et al.*) and even whole animals (Lauterbur). Some distinct changes in method are required during this progression. While some studies represented here are typical high resolution NMR analyses, others

vi

involve the use of magnetic field gradient techniques (see
also Abstracts 43–46). Approaches to aberrant growth
(tumour) detection are obviously of great interest.

The impression which the editors of this book wish to
leave with the reader is of the exciting possibilities
that have been opened up by the rapid progress in NMR
methods in the last 5 to 10 years. We are left with few
final answers but with many questions about biological
systems. (i) Should proteins be reclassified on the basis
of their motility as described by NMR? (ii) What role does
motility of side-chains play in enzyme mechanism?
(iii) Can we see how information is transferred within a
protein matrix? (iv) How do proteins interact with one another and with RNA and DNA? (v) What is the nature of
biological organisation in membranes and organelles?
(vi) How much new information can be obtained about whole
tissues by applying NMR methods? (vii) Will NMR become a
method of value in medicine? It will take another 5 to 10
years for the answers to these questions to emerge.

Oxford
June 1977

I.D.C.
R.A.D.
R.E.R.
R.J.P.W.

ACKNOWLEDGEMENTS

The successful financial outcome of the British Biophysical
Society's Spring Meeting on NMR in Biology, of which this
book is a record, is because of the extremely generous support we received from our sponsors. Their generosity has
to be seen too in the context of the difficult economic
climate prevailing. The sponsors are: Academic Press,
Bankside Manufacturing Company, Banquets (Oxford), Body-
cote International, Bruker (U.K.), Imperial Chemical Industries, Nicolet Instruments, Oxford Instrument Company,
Oxford University Press, Pergamon Press, The Royal Society,
Varian Associates, Wellcome Reagents, The Wellcome Trust.

We also wish to thank the untiring and enthusiastic
efforts of our local helpers including S.K. Dower, P.
Gettins, R. Jackson, G. Moore, Arabella Morris, B. Sutton,
S. Wain-Hobson, K. Willan.

The time from receipt of the last manuscript to actual
publication was less than 100 days. The co-operation of
the authors and Judy Sibley of Academic Press was outstanding, as was that of Mrs. Anne Joshua who patiently
and carefully typed the entire manuscript.

CONTENTS

CONTENTS

CONTENTS

CONTENTS

CONTENTS

CONTENTS

CONTENTS

CONTENTS

STRUCTURAL BASIS OF THE ACTIVATION
AND ACTION OF TRYPSIN

ROBERT HUBER and WOLFRAM BODE

Max-Planck-Institut für Biochemie,
D-8033 Martinsried bei München, Germany

Introduction

A hundred years ago, in 1876, W. Kühne first introduced
the word 'Enzym' to describe the pancreatic protease trypsin [2,3]. Trypsin turned out to be an enzyme of utmost
importance. It is a prominent member of a whole family of
functionally and structurally related digestive enzymes.
Among these are chymotrypsin and elastase, which have also
been analysed in their three-dimensional structure [4,5].

Trypsin forms obviously the functional principle of
very large and highly specific proteases involved in blood
clotting [6] and complement binding [7]. Amino acid sequence studies of some of these proteases show the presence of a trypsin-like core with large segments attached
to it which modify the specificity and are responsible for
interaction with other macromolecules of the system [8].
Most of the individual steps in the cascade reactions leading to blood clotting or complement binding are specific
proteolytic cleavages liberating and activating yet another protease. Also, in this respect trypsin is a prototype as it is biosynthesized as an inactive precursor,
trypsinogen, which is activated by limited proteolysis
[9]. Limited proteolysis in order to induce enzymatic
activity is indeed a most important regulatory phenomenon.
Recently, it was found that it is also involved in phage

2 R. HUBER and W. BODE

maturation [10].

A recent, quite fascinating finding was, that some of the naturally occurring protease inhibitors are liberated from inactive precursors (pro-inhibitors) by limited proteolysis [11]. This indicates the existence of a regulatory hypercycle involving proteases and their inhibitors: Proteases activate, but also destroy (by proteolytic action on temporary inhibitors [12]) their inhibitors. A further, quite common effector in protease action is calcium, possibly by virtue of stabilizing the three-dimensional structure [13]. Trypsin has a well defined calcium binding site [14] and calcium influences its functional and structural properties. Trypsin is indeed a prototype and detailed understanding of its structure and function is of general relevance.

In this manuscript we will describe the results and implications on functional properties of our crystal structure studies of trypsinogen, trypsin, PTI* trypsin inhibitor and their complexes. The complex formed by STI trypsin inhibitor and trypsin has been analysed crystallographically and some of its structural features will be discussed [15]. The crystal structure of DIP inhibited trypsin has been determined independently [16].

The structure and mechanism of the closely related protease chymotrypsin has recently been lucidly described in an account by Blow [17]. We will therefore focus on the structural basis of the activation mechanism of trypsin and on differences and different views of the catalytic mechanism.

---

*Abbreviations used: PTI, pancreatic trypsin inhibitor (Kunitz); STI, soybean trypsin inhibitor; DIP, diisopropylphosphoryl; r.m.s., root mean square; K_S, dissociation constant of the enzyme substrate complex; conformation angles are defined according to IUPAC-IUB [1]; residues of the inhibitor are indicated by (I) or italicized.

Crystallographic Refinement

We have determined various structures of the bovine trypsin family and refined at the highest resolution allowed by the crystalline order ranging from 1.9 to 1.5 Å: Trypsin (in its benzamidine inhibited and free forms at pH 8 and pH 5 [14,18]), trypsin inhibitor PTI [19], trypsinogen [20], trypsin inhibitor-trypsin complex [21,18], trypsin inhibitor-anhydro-trypsin complex [22], trypsin inhibitor-trypsinogen complex [23,20], trypsin inhibitor-trypsinogen complex + the dipeptide Ile-Val [23,20]. These were four different crystal structures and their various isomorphous variants providing a quite detailed but of course static view of trypsin activation and action.

The crystallographic refinement procedure applied improves phases and model in a cyclic process relying solely on measured diffraction intensities which can be determined rather accurately [19,21]. As the resolution even at 1.5 Å is not atomic, certain constraints in the model are retained in the refinement procedure, in particular bond lengths and certain inter-bond angles are not varied, but set to their standard values obtained from accurate small molecule structure analyses [24]. The parameters selected to be variable are those which produce the largest model variations and are least constrained. These are the main chain angles ψ, φ, τ, ω and the side chain dihedral angles χ. The pyramidalisation at main chain atoms was not varied except at one particular carbonyl carbon atom to be discussed later. Here the deviation from planarity was very large, but consistent using different data sets. τ and ω are parameters indicating conformation strain. We are aware of the improbability that, in the real molecule, conformational strain is localized only in these bond angles. Rather it tends to be distributed over several bond angles and bonds in the vicinity. Individual atomic radii, equivalent to individual 'temperature factors', were determined in most of these structure analyses as were occupa-

tion numbers of localized solvent. It should be emphasized, that resolution even at 1.5 Å is insufficient to
decompose the 'temperature factor' in thermal vibration
and statistical disorder. There are measurements based on
Mössbauer spectroscopy with myoglobin crystals indicating
that about two thirds of the 'temperature factor' is due
to thermal vibration [25]. It was remarkable to find by
NMR measurements that even tightly packed, internal aromatic side chains in PTI flip by 180° with a considerable
frequency (a few thousand times per second) [26-28] indicating breathing motions of the inhibitor peptide of large
amplitudes at room temperature. X-ray models represent
the time averaged equilibrium structure and conjure up a
static picture which the real molecules do not have.

   The average 'temperature factor' of trypsin, trypsinogen and PTI crystals is around 10 Å, but around 20 Å for
the various complex species. These are values characteristic for highly and medium ordered crystals respectively.
We believe that in complex crystals the statistical disorder of the molecules as a whole is increased as a consequence of less tight packing in the crystals.

   The accuracy of the refined models is estimated to about
0.1 Å from crystallographic considerations and, most
objectively, from a comparison of molecular models obtained
from different crystal structures as trypsin in its free
state and complexed with inhibitor etc. This accuracy is
considerably better (by a factor of 5) than obtained from
analyses based on isomorphous phases and essential for
reliable statements on hydrogen bonds, distortions etc.
In addition refinement provides an objective indication
of flexibility, segmental disorder, partial occupation etc.
Fig. 1 should give an impression of improvement in quality
of the electron density map obtained by structure refinement of the inhibitor structure [19].

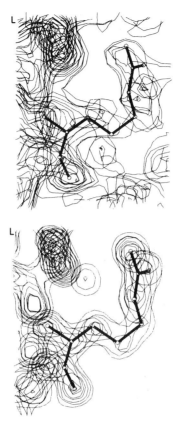

Fig. 1. Electron density and model of PTI aroung the Arg 53(I) side chain at 1.9 Å resolution based on isomorphous phases (a) and at 1.5 Å resolution based on refined phases (b).

Trypsin and trypsinogen structure

The amino acid residues of trypsin fold into a globular molecule. The secondary structure is characterized by predominance of the β-structure, but little helix. The β-structure is organized in two barrels. Such a barrel type folding has, since its discovery in chymotrypsin [4], been detected in several other unrelated proteins: immunoglobulin domains [29], thermolysin [30], alcohol dehydrogenase [31]. The chain order varies in these various barrel structures, indicating different folding pathways, but a thermodynamically favourable structural principle. The

interior of the barrels is packed with hydrophobic amino
acid side chains, as densely as observed in organic crystals [32,33]. There are some hollows in trypsin filled
with structure water molecules. These water molecules are
integral constituents of the molecular structure and have
been found in free trypsin [14], trypsin inhibitor-trypsin
complex [21,18], trypsinogen [20] and also in closely
similar positions in chymotrypsin [34].

In general these water molecules are tetrahedrally co-
ordinated accepting and donating two hydrogen bonds. They
occasionally form chains and clusters. A representative
example of such a cluster is the water structure around
the bound calcium (Fig. 2) in trypsin. The pro-enzyme

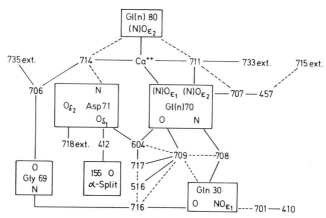

Fig. 2. Water cluster around the bound calcium in trypsin. Water molecules are indicated by numbers. Ext indicates water molecules in contact with bulk water.

trypsinogen and the active trypsin are identically folded
in about 85 per cent of the chain. The deviation of main
chain atoms in this part is 0.2 Å, insignificant with
respect of the error level of both analyses. Trypsinogen
and trypsin have been analysed in different non-isomorphous
crystal forms. It would be interesting to know whether the
observed deviation of 0.2 Å is indeed due to error in the
analysis or reflects the activation process or the different molecular packing. A clue to this might come from the

analysis of a trigonal form of trypsin which is isomorphous to trypsinogen [20] but has not yet been completely
refined. This deviation must also be discussed in relation
to thermal vibration and disorder in the crystalline state.
The refined average 'temperature factor' of trypsin and
trypsinogen crystals is around 10 $Å^2$, corresponding to a
r.m.s. vibration amplitude of 0.35 Å. Assuming that a large
part of this is indeed thermal vibration (predominantly
oscillation of rigid groups around dihedral angles, inter-
bond angle bending and deformation of planar peptide
groups, which cost little energy) we realize that structural differences of 0.2 to 0.3 Å should not significantly
alter functional properties of proteins. This is in accord
with molecular orbital calculations which show flat potential wells for nucleophile/electrophile addition [35] or
for the catalytic events in serine proteases [36] as characteristic examples. There are some large conformational
changes of external amino acid side chains in trypsinogen
compared to trypsin, probably due to crystal lattice packing requirements. These residues do not seem to be involved in activity.

Segmental Flexibility in Trypsinogen

What is the structural basis of the inactivity of
trypsinogen? Activation is brought about by the cleavage
of the N-terminal activation hexa-peptide [9]. This leaves
the majority of the molecule unchanged but drastically
alters the conformation of 4 segments tightly interdigita-
ting in trypsin: the N-terminus to Gly 19, Gly 142 to Pro
152, Gly 184 to Gly 193, Gly 216 to Asn 223.

These segments are indicated in Fig. 3. For brevity we
call these segments the 'activation domain'. There is no
electron density for these segments in the Fourier map of
trypsinogen (Fig. 4), because they are either flexibly
waggling or adopting (at least 3) different conformations.
The latter deduction stems from significance considerations

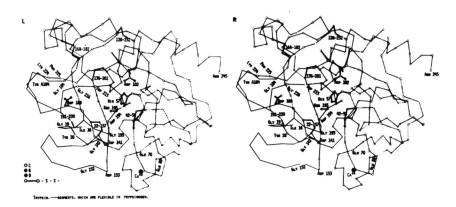

Fig. 3. Stereo drawing of the C^{α} carbon positions in trypsin. Residues linked by single lines are flexible in trypsinogen. Residues linked by double lines are fixed in trypsinogen. The catalytic residues as well as the hinge residues are indicated.

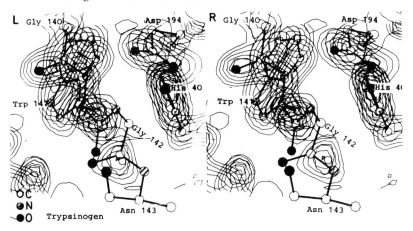

Fig. 4. Stereo drawing of the Fourier map of trypsinogen around Trp 141, Asp 194, His 40, Trp 141 - Gly 142 is a hinge. There is no density after Gly 142 N.

of the electron density map. Both situations would produce a similar effect on the appearance of the electron density, either smearing it out over a large volume or distributing it on several sites, but with low weighting. A decision as to which situation occurs might be provided by spectroscopic methods. This has not yet been attempted. Also a

crystallographic analysis at very low temperature might
provide a clue. It is remarkable that in five of the seven
hinges a glycine residue is located. Flexibility starts
rather abruptly in single residues. These glycines are
conserved in serine proteases, suggesting similar structural transitions upon activation. Glycine, which is a
residue not having a side chain, is a preferred candidate
to mediate flexibility. Immunoglobulin G molecules have
several hinges characterised by the presence of glycine
residues [37]. Three of the seven hinges in trypsinogen
have an aromatic residue adjacent to the glycine. This
residue is fixed and might serve as an anchor. The segments
around 190 and 220 are part of the pocket binding the
specificity side chain of the substrate. They are connected by the disulphide 191—220 which is also flexible. This
observation of a cross-linked, flexible loop is contrary to
the common view that disulphide linkages make structures
rigid. The triggering event leading to the formation of
the rigid, correctly designed specificity pocket is the
drastic conformational change of Asp 194 forming a link
to His 40 in trypsinogen and an internal salt bridge to
Ile 16, the newly formed N-terminus, in trypsin (Fig. 5a,b).
This structural change has first been found in chymotryp-
sinogen [38], but further structural interpretation of the
zymogen-enzyme transition disagrees substantially from what
we describe for trypsinogen. In particular, Kraut and co-
workers are able to follow the chain segments in chymo-
trypsinogen which are flexible in trypsinogen. In chymo-
trypsinogen these chain segments are in a conformation
differing from chymotrypsin.

The flexible chain segments are a tightly interdigita-
ting structural unit in trypsin as shown in the hydrogen-
bonding diagram (Fig. 6). There are more than twenty hydrogen bonds cross-linking the segments of the activation domain in trypsin which are lost or, better, replaced by
bonds to water upon mobilization in trypsinogen. Particu-

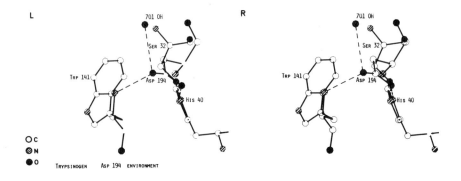

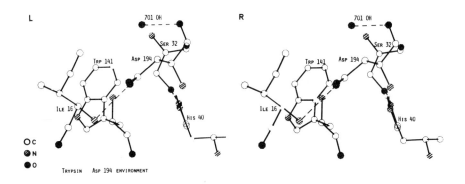

Fig. 5. Stereo drawing of the Asp 194 conformation and environment in
trypsinogen (a) and trypsin (b).

larly noteworthy is the hydrogen bonding network of the
Asp 194 carboxylate - Ile 16 ammonium ion pair, which
appears to be a clamp of the whole hydrogen bonding system.
In addition there are strong and specific hydrophobic interactions in particular of the Ile-Val N-terminus in
trypsin. The hydrogen bonding of the activation domain to
the rest of the molecule is weak consisting of only three
linkages. The activation domain appears as a rather separate entity.

The N-terminus Ile 16 occupies a pocket in trypsin
forming a salt link and several hydrogen bonds to other
residues of the activation domain and to structure water

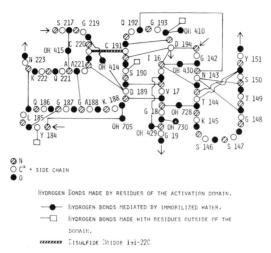

Fig. 6. The activation domain in trypsin. ———— hydrogen bonds between residues of the activation domain, ————□ hydrogen bonds to residues outside the domain.

associated with the activation domain (Fig. 7). Consequently the buried Ile N-terminus has a high pK_a of 10 [39,46]. Titration of this group leads to inactivation and possibly to a species resembling trypsinogen. This suggests a conformational linkage between the Ile 16 — and

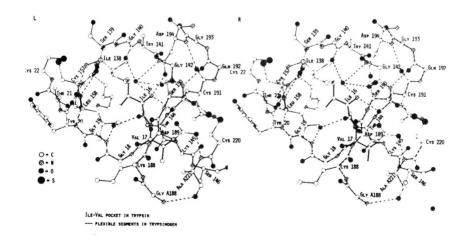

Fig. 7. Stereo drawing of the Ile 16 pocket in trypsin. Residues, which are flexible in trypsinogen are indicated ——————.

the specificity pocket: formation of one induces the formation of the other. Indeed, it is possible to demonstrate
such a linkage rigorously by inducing the structural transition from both sides. Sufficiently strong binding of a
specific inhibitor to trypsinogen rigidifies the specificity pocket *and* the Ile 16 pocket, although there is no
Ile 16 N-terminus. Such a species is observed in the trypsinogen–PTI complex. The rigidification of the Ile 16 binding pocket with the concomitant formation of the specificity pocket is also possible by reacting trypsinogen with
a peptide sequentially related to the trypsin N-terminus
i.e. Ile-Val, according to spectroscopic evidence [23].
This interaction is highly specific. Even the closed related Val-Val peptide is 30 fold less active. The peptide
Ile-Ala is inactive. But the presence of a Ile-Val dipeptide alone is not sufficient. In addition, a strong,
specific inhibitor, p-guanidobenzoate covalently bound to
trypsinogen is required to bring about the structural
transition [23].

The activity of the 'foreign' Ile-Val peptide to induce
the trypsinogen–trypsin structural transition is therefore
far inferior to the 'own' Ile-Val N-terminus. This is due
to a lower effective concentration (we compare a bi-
molecular with a mono-molecular reaction) but also probably due to conformational constraints of the 'own' N-
terminus favouring the binding conformation. Similar
effects are discussed for rate and binding enhancement in
intramolecular and enzymatic reactions [40–42]. The trypsinogen–PTI complex has been analysed crystallographically
in detail. Here, the empty Ile-Val binding pocket is
filled with some structure water which can readily be displaced by an Ile-Val dipeptide. The dipeptide binds with
about 8 kcal/mol to the complex [43]. Also the Ile-Val
addition compound has been analysed crystallographically
and shows a positioning of the Ile-Val peptide identical
to the N-terminus in the trypsin PTI complex. In the

trypsinogen–PTI complex the activation domain is structured
except the N-terminus which remains mobile to Gly 19.
Binding of the Ile-Val peptide leads to a stronger fixation
of the 142–152 loop indicated by reduced temperature factors (27 $Å^2$ and 17 $Å^2$ respectively, corresponding to r.m.s.
vibration amplitudes of 0.58 Å and 0.46 Å for the complexes
without and with dipeptide bound). Structurally the trypsinogen–PTI complex is very similar to the trypsin–PTI complex with a main chain deviation of 0.15 Å. Also the Lys
15 (I) C tetrahedral distortion is identical within the
limits of error, as well as the catalytic residues (deviation 0.18 Å).

Trypsinogen exhibits a new fascinating facet as an
allosteric enzyme. We may depict a thermodynamic cycle
linking the various structures and their equilibria (Fig.
8).

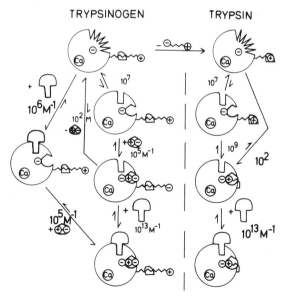

Fig. 8. Simplified equilibrium scheme for the trypsinogen, PTI, Ile-
Val system (left) and the trypsin, PTI system (right) in 0.1 M Tris/
HCL, 0.02 M CaCl at pH 8.0. Equilibrium constants are measured
(large) or estimated (small). Ⅶ flexible segments, ⊕ ⊖ Ile-Val
dipeptide.

Apart from minor structural variations in amino acid
side chain conformations and the different degrees of
structuring of the 142–152 loop there are only two different, interconvertible structures, trypsinogen and trypsin,
representing a two-state allosteric model. The various
species observed crystallographically are A, B, C, D and
E. Equilibrium constants are experimentally determined
(large numbers) [23,43–46] or have been inferred (small
numbers) from these measurements assuming that the equilibria are identical for species with the same structural
features. The equilibrium constant for H ⇌ B was taken
from reference [46]. Species F, G, H and I are hypothetical, but have to be postulated as intermediates in the
thermodynamic scheme. They are unstable with respect to
species A, B, C, D and E. The allosteric properties are
mediated by the segmental flexible-rigid transition of the
activation domain discussed. There are indications, that
a smaller effect is operative in immunoglobulins [37].
This type of signal transfer is different from what is
seen in the oligomeric haemoglobin [47]. Here the allosteric effects are mediated by a change in quarternary
structure. The flexible-rigid transition might be a general phenomenon in single subunit proteins. In anthropo-
morphic terms it is more difficult for nature to develop
a variant and stable folding, obstructing substrate binding, than to leave parts of the molecule unfolded at all.
This is a quite natural situation in the course of protein
folding itself, which is believed to occur domain-wise.

It is unclear, whether the allosteric induction of
activity in trypsinogen and related zymogens without activation peptide cleavage is of physiological significance.
There are observations of the development of proteolytic
activity prior to activation peptide cleavage in the plas-
minogen streptokinase complex [48]. Similar phenomena
appear to occur in the complement system [7].

Does the foregoing provide an answer to the question

why trypsinogen is inactive? The activation domain in
trypsin plays an essential role in specific binding of the
substrate as will be described later. This domain is flexible in trypsinogen and unable to mediate substrate binding.
The Gibbs free energy for the conformational transition to
trypsin that leads to rigidification of the substrate
binding segments in trypsinogen is estimated to be 9 kcal/
mol from the difference in binding energy of PTI to trypsin and trypsinogen [44,45]. K_S for substrate binding to
trypsinogen should be reduced correspondingly. As K_S for
good, specific substrates is around 6 cal/mol [49], trypsinogen should be essentially inactive.

The catalytic residues to be described later have identical conformations in trypsin and trypsinogen, except a
slight main chain shift and a χ rotation of Ser 195 (Fig.
9). But this dihedral angle appears to be somewhat variable

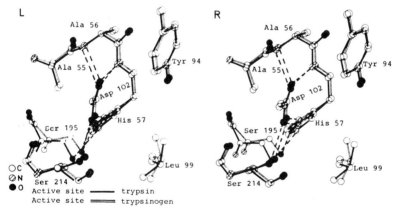

Fig. 9. Stereo drawing of the catalytic residues in trypsinogen
(══════) and trypsin (▬▬▬▬) overlaid.

as it is slightly different also in trypsin at different
pH values. Furthermore, we discussed that small structural
variations (a few tenths of an Angström) are unlikely to
change functional properties fundamentally. Such structural features agree with the observation of some (very low)
activity of trypsinogen against specific substrates but
high activity against non-specific substrates [50]. The

catalytic inefficiency of trypsinogen is predominantly expressed in low values of K_S.

Inhibitor Structure and Action

The basic pancreatic trypsin inhibitor (PTI) is a small,
pear-shaped protein molecule. [19]. Its structural motif
recurs in many rather different molecular species, as deduced from amino acid sequence homologies of various inhibitors. Even a duplicated molecule consisting of two PTI-
like domains in one polypeptide chain has been detected
[51]. PTI is an extremely potent inhibitor of trypsin and
other proteases. It binds to trypsin with an association
constant of 10^{13} M^{-1} which is the highest value known for
protein-protein interactions [44]. It was of fundamental
importance to find with other natural inhibitors [52], and
recently also with PTI [53], that interaction with proteases involves formation of a modified inhibitor species
which has the active site peptide bond cleaved. The actual
experiment is to incubate inhibitor with catalytic amounts
of protease and to analyse the reaction product which consists of an equilibrium mixture of modified and virgin inhibitor. Modified inhibitor can be isolated and incubated
with protease to yield the same equilibrium composition as
in the forward reaction. This tells us, in general, that
inhibitor protease interaction involves catalytic action
and the complex is an intermediate in catalytic peptide
bond cleavage.

The kinetic reaction scheme and the thermodynamic data
of PTI-chymotrypsin and -trypsin interaction have been
thoroughly studied in detail [54–56].

There is kinetic evidence for a minimal mechanism

$$E + I \rightleftharpoons L \rightleftharpoons C \rightleftharpoons L^* \rightleftharpoons E + I^*$$

where E is enzyme, I, inhibitor, L, loose complex, C stable
complex and * identifies the inhibitor species with the

active site peptide cleaved. As there is little doubt that
peptide bond hydrolysis involves an acyl-enzyme species
[57], the equation right to C has to be expanded including
the acyl intermediate but there is no definitive kinetic
evidence for it. Formation of L is a very fast, entropy-
driven reaction, while the conversion of L to C is slow
and has also a favourable enthalpy contribution [54,55].
The pH dependence of the association rate closely resembles
that observed for the rate of catalysis of serine proteases [57,58]. Dissociation rates C → L (and also C → L*,
as measured for soy-bean trypsin inhibitor [55]) are extremely slow at neutral pH. Quite generally, PTI and other
natural inhibitors have the characteristics of excellent
substrates in the association step, but inhibit by virtue
of their slow dissociation. We may safely assume that the
structural features in the PTI—trypsin complex are relevant for trypsin substrate interaction.

The Complex

Only a small proportion of both molecules is in contact
in the complex: 14 amino acid residues out of 58 of the
inhibitor, and 24 amino acid residues out of 224 of trypsin. The contact is characterized by a complicated network
of hydrogen bonds and a large number of van der Waals contacts. The contact is tightly packed with a density identical to that observed in the interior of protein molecules
[33]. A dominant interaction is made by the specificity
side chain of PTI (Lys 15(I)) inserted into the specificity
pocket (Fig. 10). The specific interaction occurs between
the positively charged Lys 15 (I) ammonium group of the inhibitor and the Asp 189 carboxylate of the enzyme. This
carboxylate is responsible for the primary specificity of
trypsin for positively charged side chains. N^ζ of Lys 15
uses fully its hydrogen bonding capabilities, donating
three hydrogen bonds to the carbonyl oxygen of Ser 190 and
two water molecules 416 and 414, one of which is bonded to

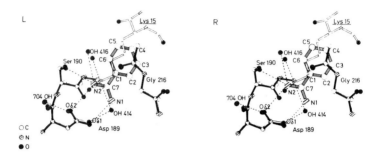

Fig. 10. Stereo drawing of the Lys 15 (I) binding in the specificity pocket of trypsin as seen in the PTI–trypsin complex (═════) overlaid with benzamidine (═════) as seen in benzamidine inhibited trypsin (▬▬▬▬).

Asp 189. The binding of a small inhibitor molecule benzamidine to trypsin is depicted in Fig. 10 as a close look at
the specificity pocket of benzamidine inhibited trypsin.
Benzamidine is a small synthetic inhibitor simulating
arginine residues. It forms hydrogen bonds directly to the
Asp 189 carboxylate, to Ser 190 O^{γ} to a water molecule 416
and Gly 219 O. There is a small structural rearrangement
of the specificity pocket to fulfil the different hydrogen
bonding requirement of an ammonium group (lysine) and a
guanidinium group (benzamidine, arginine). Water 414 is
expelled, when benzamidine binds, and Ser 190 O is attrac-
ted by lysine and the peptide plane slightly rotated (Fig.
10). Comparing free trypsin with benzamidine inhibited
trypsin (Fig. 11) we find only small structural changes,
the most conspicuous being the expulsion of one molecule

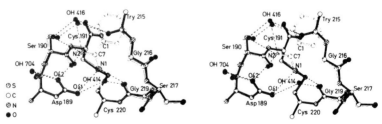

Fig. 11. Stereo drawing of the specificity pocket of free trypsin (▬▬▬▬) with benzamidine inhibited trypsin overlaid (═════).

of structure water. It is clear that in free trypsin the
specificity pocket must be filled with some further water
molecules which are, however, mobile and do not contribute
to the electron density of the crystal structure. These
must also be displaced upon inhibitor binding.

The specificity side chain held in the specificity pocket is one anchor to fix the inhibitor or substrate at the
enzyme surface. A second important binding interaction is
between the substrate polypeptide chain and the enzyme.
These hydrogen bonding interactions are schematically represented in Fig. 12. The substrate main chain from P_3 to
P_2' is linked via 6 hydrogen bonds to the enzyme. The enzymatic split occurs between P_1 and P_1'. An essential interaction appears to be the bifurcated hydrogen bond of P1 CO
with the NH groups of Gly 193 and Ser 195, which form the

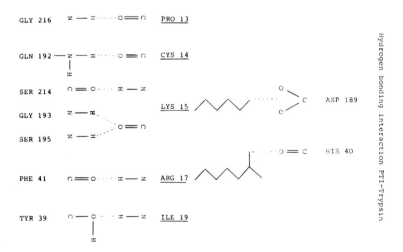

Fig. 12. Schematic hydrogen bonding interactions formed between PTI
and trypsin as seen in the complex.

oxy-anion binding hole out of reasons which will become
obvious later. This particular bond, together with the side
chain interaction in the specificity pocket, precisely
orients the scissile peptide group with respect to the catalytic residues. It should be recalled that neither the
specificity pocket nor the oxy-anion binding hole is formed
in trypsinogen due to mobility of the segments involved.

The arrangements of the catalytic triad Asp 102, His
57 and Ser 195 of these catalytic residues was first observed in chymotrypsin and their essence has been lucidly
described in a recent account [17], so that we may be
allowed to concentrate on these points where our structural studies might help in clarification.

pH Dependence

Trypsin activity as well as the rate of inhibitor association depends critically on pH (pK_a around 7) [54–56].
The group responsible was believed to be His 57, but both
NMR studies and chemical modification experiments are con-
troversial in assigning a pK_a around 7 to His 57 or Asp
102 [59–63]. Our structural studies show that free trypsin
at pH 5 and pH 8 are extremely similar. Fig. 13 shows the
catalytic site residues overlaid. There is virtually no
structural variation, except a slight Ser 195 O^γ rotation
which improves the His $N^\varepsilon - O^\gamma$ hydrogen bond in the low pH

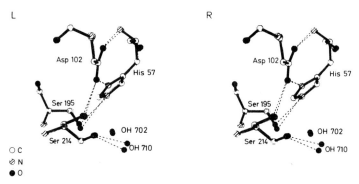

Fig. 13. Stereo drawing of the catalytic residues as seen in trypsin
at pH 5 (━━━) and pH 8 (═══) overlaid.

form. (This hydrogen bond is 3.26   and rather bent at pH 8, while it is 2.90 Å and more linear at pH 5). A second change is in the active site associated solvent 702, which is characterized by a broad, smeared-out density. The integrated electron density corresponding to this solvent is 30e at pH 5 and 18e at pH 8. Its position is 3.75 Å from Ser 195 O^γ at pH 5 and 3.08 Å at pH 8. This is compatible with the presence of a disordered sulphate ion at low pH, which is lost to a large extent and/or replaced by water at high pH. Tulinski and Wright presented evidence for this sulphate in chymotrypsin crystals at low pH [64]. This observation indicates a change in charge of the catalytic groups and prefers a positively charged His 57 at low pH. Whichever residue is protonated at low pH, its effect is felt outside the molecule and the polarity of Ser 195 O^γ, the hydrogen bonding partner of the solvent, is influenced. It is conceivable that protonation of the catalytic system reduces the nucleophilicity of Ser 195 O^γ considerably. Further complication is brought into the puzzle through the observation that the pH dependence of the anhydrotrypsin–PTI association rate is similar to the trypsin–PTI association rate except a pK_a change of 0.6 units [65]. Anhydro-trypsin has no Ser 195 O^γ and so the nucleophilicity of Ser 195 O^γ is excluded as the source of the pH dependence. We believe that sites far from the catalytic residues should also be considered. A possible candidate, under others, is the bound calcium which is lost to a large extent at pH 5 accompanied by disorder of the peptide segment providing the Ca^{++} ligands. Regulation by order-disorder would be reminiscent of the mechanism discussed for trypsinogen–trypsin.

Ser 195 O^γ position

A second aspect, where we prefer a different view from that described in Blow's account [17], concerns the Ser 195 O^γ position in the free enzyme. Trypsin at pH 5 and pH 8 is characterized by a Ser 195 O^γ dihedral angle (χ^1) of -95° and -60° respectively. As this angle is -83° in the PTI trypsin complex (tetrahedral adduct) [21] or -64° in the acyl-enzyme [34] little conformational change is required to proceed in the catalytic reaction steps (Fig. 14). In contrast, in α-chymotrypsin crystals at pH 4.5 a χ^1 angle of 93° has been observed [34]. Such a conformation requires a major conformational change to reach the tetrahedral adduct. We believe that this is an inactive conformation either stabilized by the low pH or the dimer formation in chymotrypsin crystals.

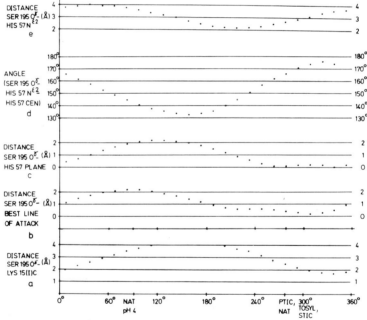

Fig. 14. Ser 195 O^γ is rotated around χ^1 and various angles and distances calculated based on the complex coordinates. NAT, pH 4 is χ^1 as observed in chymotrypsin [34], PTI C as in complex, NAT, benzamidine inhibited trypsin, STIC, STI-trypsin complex [15], TOS, tosylchymotrypsin [34]. Curves c, d, e characterize the Ser 195 O^γ His 57 N^ε hydrogen bonding, curve a and b the O^γ–C Lys 15 (I) interaction.

Arrangement of scissile peptide and catalytic residues

We will now focus on the relative arrangement of the scissile peptide group and catalytic residues as seen in the complex, shown in Fig. 15. The nucleophilic Ser 195 O^γ is approximately perpendicular to the Lys 15I-Ala 16I

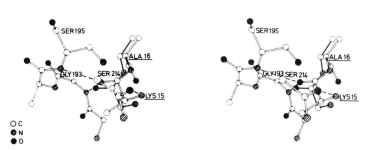

Fig. 15a. Stereo drawing of the oxy-anion binding hole as seen in the complex (═══). Overlaid is the inhibitor part in the conformation of the free inhibitor optimally fitted to the complexed inhibitor (▬▬▬). This conformation would be sterically impossible.

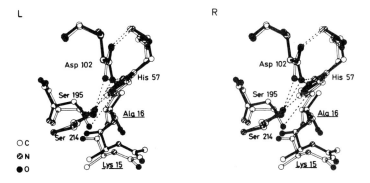

Fig. 15b. Stereodrawing of the catalytic residues as seen in the complex (▬▬▬) and in free trypsin (═══) overlaid. The inhibitor segment deduced from the free inhibitor conformation is also shown (═══). Ser 195 is pushed out of the way and His 57 seems to follow improving the hydrogen bond.

peptide plane above the carbonyl carbon. This is close to the position expected for a nucleophilic addition reaction as shown by Bürgi, Dunitz and Shefter [66]. The best line of attack would be along the tetrahedral bonding direction of C. The peptide carbonyl carbon is tetrahedrally distor-

ted characterized by an out-of-plane bend (the plane defined by C^α, C, N) of the oxygen by -34° i.e. the carbon
is half way between a trigonal and a tetrahedral conformation. The distance between Ser 195 O^γ and Lys 15 C is 2.6
Å, longer than expected for a covalent bond but shorter
than a van der Waals distance. The Lys 15(I) C—O distance
should be lengthened as well as C—N (Ala 16(I)), but the
resolution of the analysis is insufficient to confirm
this. We regard this as an intermediate state of the nucleophilic addition reaction frozen by constraints imposed
by enzyme and inhibitor. Similar phenomena had been observed in small molecule crystal structures [66], but the
tetrahedral distortion in the complex is much larger than
expected from the long O^γ ---- C distance. Theoretical
calculations of the reaction coordinate of nucleophile
electrophile addition show that the energy of the system
changes little in the range of 3 to 2 Å distance between nucleophile and electrophile [35], like anhydro-trypsin [67]
lacking SER 195 O^γ but binds the inhibitor nearly as strongly
as native trypsin [65]. In anhydro-trypsin Ser 195 has been
converted to a dehydro-alanine by a chemical reaction which
inactivates the enzyme completely. The structure analysis
of the anhydro-trypsin-PTI-complex [22] clearly revealed
the loss of Ser 195 O^γ but indicated that the pyramidalisation of the Lys 15(I) carbonyl carbon is identical to the
native complex. This distortion is obviously brought about
by the interaction of the Lys 15(I) oxygen with the oxy-
anion binding hole. Such distortion favours formation of
the tetrahedral adduct sterically and electronically. The
negative charge developing on the Lys 15(I) carbonyl oxygen is districuted through the two hydrogen bonds to Gly
195 N and Ser 195 N. It is also noticeable that the torsion angle ω (C_i - C_4 - N_{i+1} - C_{i+1}) of the Lys 15(I)—Ala
16(I) peptide group deviates by about 20° from the ideal
value of 180° in free and complexed inhibitor. Such a distortion might also assist the formation of the tetrahedral

adduct. In general, the arrangement of the scissile peptide and the catalytic groups is designed to allow the
minimum energy Ser 195 O^γ --- C approach i.e. close to the
best line of nucleophile electrophile attack and preserv-
ing the His 57–Ser 195 O^γ hydrogen bond. This is demonstrated in the model building experiment shown in Fig. 14. Ser
195 O^γ is rotated and the various distances and angles
calculated based on the trypsin–PTI complex coordinates.
During O^γ --- C bond formation (3 Å to 1.5 Å) these cri-
teria are met.

Complex formation is accompanied by slight distortions
in the inhibitor binding segment other than the pyramidalisation of the carbonyl carbon of Lys 15(I). Comparing
the main chain dihedral angles of free and complexed inhibitor the mean deviation in internal segments is around
5° while it amounts to 35° in the contact area. The inhibitor adapts to the enzyme. It would not fit to the enzyme
in its native conformation (Fig. 15a,b). There is also
some rearrangement of the Ser 195 main chain of the enzyme
upon complex formation. The inhibitor appears to push Ser
195 slightly out of the way. The hydrogen bond between Ser
195 O^γ and His 57 N^ϵ improves considerably in the complex
where it is perfectly linear with a length of 2.7 Å, compared to the long and bent bond in free trypsin at pH 8.
This has functional relevance as it enhances the nucleophilicity of O^γ in the complex.

Speculating about the structural events related to the
kinetic steps, the formation of L might be primarily a
desolvation process of enzyme and inhibitor contact surfaces. This would not only involve the few rigidly bound
solvent molecules associated with the active site residues
but in particular the numerous, mobile water molecules
loosely bound to polar surface groups. This process could
provide the entropy contribution observed. In L both molecules might still not be properly aligned with considerable relative mobility. Formation of C involves the slight

conformational changes described allowing optimal fitting
of both molecules accompanied by a favourable enthalpy
contribution. In the complex the Ser 195 O^γ position is
well defined. There is no indication of a bimodal electron
density distribution. Also His 57 does not have an increased mobility as we see it in the free enzyme. The geo-
metric observed in the complex (scheme Fig. 16) uniquely

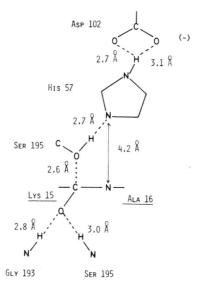

Fig. 16. Scheme of the arrangement of the catalytic residues and the
scissile peptide as seen in the complex.

defines the minimum energy pathway leading to the 'true'
tetrahedral intermediate with a covalent Ser 195 O^γ—C bond.
This requires slight further 'down' rotation of O^γ and an
'up' movement and further pyramidalisation of C. C—O and
C—N bonds are lengthened. Vibration of C along C—N leads to
to the acyl species with a planar C^α, C, O, O^γ configuration and the C—N bond broken (Fig. 17). The hydrogen is
transferred from Ser 195 O^γ to His 57 N^ε during C—O bond
formation in accord with the estimated pK_a of the groups
involved [68]. Proton transfer occurs along an existing
hydrogen bond and should be fast and kinetically insignificant. Transfer of the proton from His 57 to N Ala 16 (I)

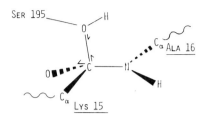

Fig. 17. Vibrational modes of C of the scissile peptide leading to the O—C covalent bond and the acyl-enzyme.

of the leaving group required before the C—N bond can break involves a conformational change of the imidazole. Its distance is more than 4 Å in the complex and it is unfavourably oriented for proton transfer (Fig. 16). A conformational change of His 57 as the rate limiting step in serine protease catalysis is discussed and in accord with experimental data [68] as well as theoretical calculations [36]. Some further crystallographic evidence is at hand about the acyl-enzyme intermediate in the catalytic pathway which has been described in Blow's account [17].

What is the structural basis of inhibition by PTI? Kinetic measurements with STI and PTI show that dissociation of the complex to starting material is extremely slow [54—56]. Dissociation to form I* is even slower in the STI case where it could be measured [55]. Dissociation to I* and I requires different mechanism. It might be slowed down by a number of reasons:

The free inhibitor might be regarded as a transition state analogue, but then it should perfectly fit the enzyme with no distortions required. This is not the case (Fig. 15), in particular looking at the characteristic pyramidalisation of C Lys 15 (I). The free inhibitor has a planar C [19,21]. But clearly these changes are small compared to a flexible substrate chain adapting to the enzyme.

The inhibitor might be strained with the strain released upon complexation. The susceptible peptide group and the adjacent peptides have indeed strained torsion angles ω, but these remain essentially unchanged in the complex [19,21].

Formation of I* might be difficult because the His 57 conformation change discussed for acyl-enzyme formation is hindered. In the PTI-trypsin complex the imidazole is indeed tightly packed and shielded from water in particular by the chain segment Gly 36 (I) to Cys 38 (I) antiparallel to the Lys 15 (I) − Ala 16 (I) chain. This chain is absent in STI, but here the bulky Tyr 62 (I) and Ile 64 (I) (equivalent to Cys 14 (I) and Ala 16 (I) in PTI) shield His 57.

C–N bond cleavage involves loss of the interactions at the leaving group side P' (Fig. 12). This is energetically unfavourable, but we do not know whether these interactions are also made with a normal substrate chain. Dissociation to starting material I requires to break the large number of favourable interactions involved. However, these are probably also made by a normal substrate chain which binds weaker.

The free inhibitor has a structure largely complementary to the enzyme. This is entropically favourable compared to a flexible substrate chain. We take this as the essential feature of an effective inhibitor.

References

1. IUPAC-IUB Commission on Biochemical Nomenclature (1970). *Biochemistry* **9**, 3471.
2. Kühne, W. (1877). *Verhandl. Naturhist.-Med. Ver. Heidelberg* **1**, 194–8.
3. Gutfreund, H. (1976). *FEBS Lett.* **62** (Supplement) E1-E12.
4. Matthews, B.W., Sigler, P.B., Henderson, R. and Blow, D.M. (1967). *Nature (London)* **225**, 811–16.
5. Shotton, D.M. and Watson, H.C. (1970). *Nature (London)* **225**, 811–16.
6. Davie, E.W. and Fujikawa, K. (1975). *Ann. Rev. Biochem.* **44**, 799–829.
7. Müller-Eberhard, H.J. (1975). *Ann. Rev. Biochem.* **44**, 697–724.
8. Magnusson, S., Sottrup-Jensen, L., Petersen, T.E., Dudek-Wojciechowska, Dudek-Wojciechowska, G. and Claeys, H. (1976). *In: Miami Winter Symposia*, Vol. 11 (D.W. Ribbons, K. Brew, eds.), Academic Press,

New York, 203–39.
9. Neurath, H. and Davie, E. (1955). *J. Biol. Chem.* **212**, 515–29.
Desneuelle, P. and Fabre, C. (1955). *Biochim. Biophys. Acta* **18**,
49–57.
10. Isobe, T., Black, L.W., Tsugita, A. (1977). *J. Mol. Biol.* **110**,
165–77.
11. Bretzel, G. and Hochstrasser, K. (1976). *Hoppe-Seyler's Z. Physiol.
Chem.* **357**, 487–9.
12. Laskowski, M., Sr. and Wu, F.C. (1953). *J. Biol. Chem.* **204**, 797–
805.
13. Kretsinger, R.H. (1976). *Ann. Rev. Biochem.* **45**, 239–66.
14. Bode, W. and Schwager, P. (1975). *J. Mol. Biol.* **98**, 693–717.
15. Sweet, R.M., Wright, H.T., Janin, J., Cothia, C.H. and Blow, D.M.
(1974). *Biochemistry* **13**, 4212–4228.
16. Stroud, R.M., Kay, L.M. and Dickerson, R.E. (1974). *J. Mol. Biol.*
83, 185–208.
17. Blow, D.M. (1976). *Acc. Chem. Res.* **9**, 145–52.
18. Bode, W., Schwager, P. and Huber, R. (1976). *In: Miami Winter
Symposia, Vol. 11, Proteolysis and Physiological Regulation* (D.W.
Ribbons, K. Brew, eds.), Academic Press, New York, 43–76.
19. Deisenhofer, J. and Steigemann, W. (1975). *Acta Cryst.* **B 31**, 238–
50.
20. Bode, W., Fehlhammer, H. and Huber, R. (1976). *J. Mol. Biol.* **106**,
325–35.
Fehlhammer, H., Bode, W. and Huber, R. (1977). *J. Mol. Biol.* **111**,
in press.
21. Huber, R., Kukla, D., Bode, W., Schwager, P., Bartels, K.,
Deisenhofer, J. and Steigemann, W. (1974). *J. Mol. Biol.* **89**, 73–
101.
Rühlmann, A., Kukla, D., Schwager, P., Bartels, K. and Huber, R.
(1973). *J. Mol. Biol.* **77**, 417–36.
22. Huber, R., Bode, W., Kukla, D., Kohl, U. and Ryan, C.A. (1975).
Biophys. Struct. Mech. **1**, 1–13.
23. Bode, W and Huber, R. (1976). *FEBS Lett.* **68**, 231–5.
24. Diamond, R. (1974). *J. Mol. Biol.* **82**, 371–91.
25. Parak, F. and Formanek, H. (1971). *Acta Cryst.* **A 27**, 573–8.
26. Gelin, B.R., Karplus, M. (1975). *Prcc. Nat. Acad. Sci. USA* **72**,
2002–6.
27. Wagner, G., De Marco, A., Wüthrich, K. (1976). *Biophys. Struct.
Mechanism.* **2**, 139–58.
28. Hetzel, R., Wüthrich, K., Deisenhofer, J. and Huber, R. (1976).
Biophys. Struct. Mechanism **2**, 159–80.
29. Davies, D.R., Padlan, E.A. and Segal, D.M. (1975). *Ann. Rev.
Biochem.* **44**, 639–67.
30. Colman, P.M., Jansonius, J.N. and Matthews, B.W. (1972). *J. Mol.
Biol.* **70**, 701–24.
31. Bränden, C.I., Jörnvall, H., Eklund, H. and Furugren, B. (1975).
In: The Enzymes (P.D. Boyer, ed.), Vol. 11, 3rd edn., pp. 103–90.
Academic Press, New York.
32. Lee, B. and Richards, F.M. (1971). *J. Mol. Biol.* **55**, 379–400.
33. Cothia, C.H. and Janin, J. (1976). *J. Mol. Biol.* **100**, 197–212.
34. Birktoft, J.J. and Blow, D.M. (1972). *J. Mol. Biol.* **68**, 187–240.
35. Bürgi, H.B., Lehn, J.M. and Wipff, G. (1974). *J. Amer. Chem. Soc.*
96, 1956–7.

36. Scheiner, S. and Lipscomb, W.N. (1976). *Proc. Nat. Acad. Sci. USA* **73**, 432–6.
37. Huber, R., Deisenhofer, J., Colman, P.M., Matsushima, M. and Palm, W. (1976). *Nature (London)* **264**, 415–20.
38. Freer, S.T., Kraut, J., Robertus, J.D., Wright, H.T. and Xuong, Ng.H. (1970). *Biochemistry* **9**, 1997–2009.
39. Fersht, A.R. (1971). *Cold Spring Harbor Symp. Quant. Biol.* **36**, 71–3.
40. Page, M.I. and Jencks, W.P. (1971). *Proc. Nat. Acad. Sci. USA* **68**, 1678–83.
41. Koshland, D.E., Jr., Carraway, K.W., Dafforn, G.A., Gass, J.D. and Storm, D.R. (1971). *Cold Spring Harbor Symp. Quant. Biol.* **36**, 13–20.
42. Bruice, T.C. (1971). *Cold Spring Harbor Symp. Quant. Biol.* **36**, 21–27.
43. W. Bode, in preparation.
44. Vincent, J.P. and Lazdunski, M. (1972). *Biochemistry* **11**, 2967–77.
45. Vincent, J.P. and Lazdunski, M. (1976). *FEBS Lett.* **63**, 240–4.
46. Hartley, B.S. and Shotton, D.M. (1971). *In: The Enzymes* (P.D. Boyer, ed.), Vol. 3, 3rd edn., Academic Press, New York, London, pp. 323–73.
    Spomer, W.E. and Wooton, J.F. (1971). *Biochim. Biophys. Acta* **235**, 164–71.
47. Perutz, M.F. (1970). *Nature (London)* **228**, 726–34.
48. Bajaj, S.P. and Castellino, F.J. (1977). *J. Biol. Chem.* **252**, 492–8.
    Summaria, L., Arzadon, L., Bernabe, P. and Robbins, K.C. (1974). *J. Biol. Chem.* **249**, 4760–9.
49. Quast, U., Engel, J., Heumann, H., Krause, G. and Steffen, E. (1974). *Biochemistry* **13**, 2512–20.
50. Kerr, M.A., Walsh, K.A. and Neurath, H. (1975). *Biochemistry* **14**, 5088–94.
51. Hochstrasser, K., Wachter, E. and Bretzel, G. (1976). *Hoppe-Seyler's Z. Physiol. Chemie* **357**, 1659–61.
52. Finkenstadt, W.R. and Laskowski, M., Jr. (1967). *J. Biol. Chem.* **242**, 771.
53. Jering, H. and Tschesche, H. (1974). *Angew. Chem., internat. edit.* **13**, 661–3.
54. Engel, J., Quast, U., Heumann, H., Krause, G. and Steffen, E. (1974). *In: Bayer Symposium V Proteinase Inhibitors* (H. Fritz, H. Tschesche, L.J. Greene, E. Truscheit, eds.), Springer Verlag, Verlin, Heidelberg, New York, 412–19.
55. Finkenstadt, W.R., Hamid, M.A., Mattis, J.A., Schrode, J., Sealock, R.W., Wang, D., and Laskowski, M. Jr. (1974). *In: Bayer Symposium V 'Proteinase Inhibitors'* (H. Fritz, H. Tschesche, L.J. Greene, E. Truscheit, eds.), Springer Verlag, Berlin, Heidelberg, New York, 389–411.
56. Lazdunski, M., Vincent, J.P., Schwertz, H., Peron-Renner, M. and Pudles, J. (1974). *In: Bayer Symposium V Proteinase Inhibitors* (H. Fritz, H. Tschesche, L.J. Greene, E. Truscheit, eds.), Springer Verlag, Berlin, Heidelberg, New York, 420–31.
57. Hess, G.P. (1971). *In: The Enzymes* (P.D. Boyer, ed.), Vol. 3, 3rd edn., pp. 213–48, Academic Press, New York.

58. Henderson, R., Wright, C.S., Hess, G.P. and low, D.M. (1972). *Cold Spring Harbor Symp. Quant. Biol.* **36**, 63–70.
59. Robillard, G. and Shulman, M.G. (1974). *J. Mol. Biol.* **86**, 519–40.
    Robillard, G. and Shulman, M.G. (1974). *J. Mol. Biol.* **86**, 541–58.
60. Hunkapiller, M.W., Smallcombe, S.H., Whitaker, D.R. and Richards, J.H. (1973). *Biochemistry* **12**, 4732–42.
61. Markley, J.L. and Porubcan, M.A. (1976). *J. Mol. Biol.* **102**, 487–509.
    Markley, J.L. and Ibanez, I.B. (1976). *Biochemistry* (submitted).
62. Köppe, R.E. and Stroud, R.M. (1976). *Biochemistry* **15**, 3450–8.
63. Fersht, A.R. and Sperling, J. (1973). *J. Mol. Biol.* **74**, 137–49.
64. Tulinsky, A. and Wright, L.H. (1973). *J. Mol. Biol.* **81**, 47–56.
65. Vincent, J.P., Peron-Renner, M., Pudles, J. and Lazdunski, M. (1974). *Biochemistry* **13**, 4205–11.
66. Bürgi, H.B., Dunitz, J.D. and Shefter, E. (1973). *J. Am. Soc.* **95**, 5065–7.
67. Ako, H., Foster, R.J. and Ryan, C.A. (1974). *Biochemistry* **13**, 132–9.
68. Satterthwait, A.C. and Jencks, W.P. (1974). *J. Amer. Chem. Soc.* **96**, 7018–31.
69. Vincent, J.P., Lazdunski, M. (1976). *FEBS Lett.* **63**, 240–4.
    Quast, U, in preparation.

[This is an extended version of a manuscript submitted to Account of Chemical Research.]

AN NMR VIEW OF PROTEIN STRUCTURE

I.D. CAMPBELL

Department of Biochemistry, Oxford

Introduction

It is apparent that a complete specification of a structure requires the chemical composition and the spatial relationships of all the building blocks to be known. In the
last 20 years this knowledge has been gained for many proteins by the painstaking efforts of the protein chemist
and the X-ray crystallographer. In a complex 'machine',
such as a protein, this information is not sufficient to
specify the structure completely; two additional pieces of
information are required, namely, the flexibility and the
energetics of the structure [1].

Spectroscopic methods, which can study particular
groups in a protein in an environment similar to *in vivo*
conditions, are particularly well suited to investigate
aspects of the structure which may be designed to be very
sensitive to environment, e.g. flexibility and energetics.
Of the spectroscopic techniques, NMR is inherently the
most powerful because of its ability to study virtually
all the atoms of the system. It is also capable of yielding quantitative information about the spatial positions,
the conformational mobility and the thermodynamics of
individual groups in a protein, i.e. it can measure in the
required coordinate system $(x,y,z,t,\Delta G)$. The ways in which
NMR can be used to give information about protein struc-

ture are briefly reviewed in this article.

The NMR Method

The technical advances in NMR over the last 12 years
have been truly remarkable. The field strengths available
have increased by a factor of four and the sensitivity of
the instruments has increased by a factor of at least 20.
These advances make it possible to resolve large numbers
of resonances from many of the amino acids using both [1]H
and ^{13}C NMR. ^{1}H NMR has been most successful in observing
resonances from methyl groups and both nitrogen and carbon
bound hydrogens on aromatic rings [2]. [13]C, on the other
hand, has been most successful in observing quaternary
carbons of aromatic groups [3] and carboxyl groups [4].
(A typical ^{1}H spectrum of a protein of molecular weight
15,000 is obtained from .02 gms of material in 15 minutes,
while a typical ^{13}C spectrum is obtained from 2 gms of
material in 20 hours.)

In spite of technical improvements, some important
problems remain. Large numbers of resonances occur in a
small frequency range and these must be *resolved* and assigned to a particular type of amino acid. No doubt the
very high field magnets (~ 11 tesla) now under construction around the world will help but other methods, such as
difference spectroscopy, mathematical manipulation and
pulse techniques are also very powerful. These methods
have been reviewed recently and will not be discussed further here [5,6].

The association of a resolved resonance with a particular amino acid, *the first stage of assignment*, can be
achieved in ^{1}H NMR by spin decoupling methods (in proteins
up to about 20,000 daltons) since the coupling pattern for
each amino acid is unique [2]. In some cases the amino
acid can be identified by titration (e.g. histidine and
tyrosine) and a powerful method is selective deuteration
[7]. In some respects the first stage of assignment is

easier for ^{13}C NMR because of the wider range of chemical shifts. Selective double resonance [8] and isotope exchange techniques can also be used as assignment aids [3].

The second stage of assignment is to relate a resonance to a particular amino acid in the sequence. This stage of assignment is usually intimately linked to knowledge of the spatial relationship of groups in the protein and is discussed below [2].

The Determination of Relative Positions by NMR

In a protein with no preferred conformation in solution, all the resonances from one chemical group in one type of amino acid are usually indistinguishable. In a globular protein with a preferred conformation in solution, the environment of each amino acid is, in general, different and each resonance may be shifted from its random coil position by up to a few parts per million (ppm). It is convenient to call these environmental shifts *secondary shifts*. In Fig. 1 the resonances of the four phenylalanine residues of horse cytochrome *c* have been identified and the secondary shifts are readily observed. This ability of NMR to observe signals from atoms of individual amino acids in a folded protein in solution is unique.

Can structural information be obtained from secondary shifts of the sort observed in Fig. 1. Unfortunately, the answer, at this time, is that the information which can be derived is limited. One of the reasons for this is that the sources of the secondary shifts are not well understood. *Aromatic rings* probably contribute most and most analyses of these effects concentrate on this source only [6,10,11]. Whether other groups, such as carbonyl groups, contribute significantly is not yet known, but it seems likely that detailed analysis of a system, such as lysozyme [6], should lead to the development of useful empirical rules in the near future.

In spite of difficulties in predicting the environment

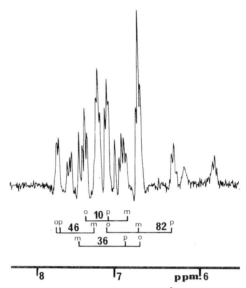

Fig. 1. The aromatic region of the 270 MHz ^{1}H spectrum of horse cyto-chrome c. The resonance positions of the four phenylalanines are shown. The ortho (o), meta (m) and para (p) proton resonances are spin coupled and the positions and multiplicity of each resonance is determined by double resonance experiments. The numbers 10, 36, 46 and 82 refer to tentative second stage assignments of the resonances. The data are taken from [9] and [15].

from the observed intrinsic secondary shift, useful infor-mation can sometimes be obtained about *structural invari-ance* using secondary shifts because the secondary shifts are very sensitive to environment.

Table 1 contains information about a tryptophan residue in various types of cytochrome c. In horse cytochrome c there is only one tryptophan, thus the second stage assign-ment problem does not exist. The approximate secondary shifts of a doublet and a triplet which arise from two non adjacent protons on the benzenoid ring of the tryptophan are listed. The shift on the triplet is very large and it should thus be very sensitive to environment. All the cyto-chromes listed, including two from bacteria and one from blue green algae, have almost identical shifts on two non adjacent protons in the benzenoid ring. This strongly suggests that a tryptophan always exists in a very similar

TABLE 1

Properties of a Tryptophan in Cytochrome c

Species	Primary sequence ...His..Met..Trp...	Secondary Shift (ppm) doublet	triplet	Ref.
Horse (c)	..18...80...59...	+.05	+1.4	12
Tuna (c)	..18...80...59...	+.05	+1.4	12
Yeast (c)	..18...80...59...	+.25	+1.4	12
R. viridis (c_2)	..17...79...58...	+.1	+1.5	12
P. mendocina (c_{551})	.16...61...56...	+.0	+1.4	12
E. gracilis (c_{552})	.14...56...59...	+.1	+1.6	13

environment, probably near to the haem in these proteins from widely divergent species. Structural homology is not obvious from the primary sequence data summarised in the table since the tryptophan is placed at widely different positions with respect to the histidine and methionine ligands to the iron.

We must now consider the problem of *second stage assignments*. For example in Fig. 1, how do we relate the primary sequence phenylalanines (10, 36, 46 and 82) to the observed phenylaline resonances? This task is, in general, perhaps the most difficult of all the problems facing the NMR spectroscopist studying proteins. The best methods relate some spectral property directly to the primary sequence, e.g. the work of Markley [14] in using differential exchange rates of C-(2) protons of histidine. Selective changes brought about by chemical modification or by using a protein with a slightly different sequence are also very valuable (e.g. phe 46 in Fig. 1) but this type of study is not always unambiguous and it seems unlikely that many resonances could be assigned in this way [6]. Most second stage assignment procedures rely on the 3D structure, determined by the X-ray crystallographer, as a good approximation to the solution structure. This allows a whole range of methods to be used; binding of ligands and paramagnetic probes, calculating the secondary shifts from aromatic rings, relating pK_a values and isotope exchange rates to environment and so on [2,5]. In the example illustrated in Fig. 1, phe 82 was assigned [15] by observing the secondary shifts on both the oxidised and reduced protein. In the oxidised form there is an additional 'pseudo contact' shift because of the paramagnetism of the iron and this is temperature dependent. This suggests that the group is the nearest of the phenylalanines to the haem and the presence of the upfield shifted para proton resonance in the reduced protein is consistent with the position of phe 82 with respect to the haem determined by Dickerson *et al.* [16] using X-ray crystallography. While this assignment is made using the X-ray structure, additional information is obtained since the NMR data on the oxidised protein are inconsistent with the large conformational change between the two oxidation states originally proposed from the X-ray work.

It is clear that if second stage assignments can be made, a considerable amount of information is available about local environmental effects on the assigned group. To quantify properly the spatial relationships of these groups, however, one must introduce controlled extrinsic perturbations to the system. *Paramagnetic metal ions*, especially the lanthanide ion series, are particularly well suited for this, for while the binding properties of the members of the series are similar, the magnetic properties vary widely [17,18]. The observed induced shifts and line broadening observed give a direct measure of the relative distance of various groups to the metal ion-binding site. In the particular case of lysozyme [6] this technique has led to a very detailed view of the structure of this protein in solution. Perhaps the main limitation of the method, after the assignment problem which is a

general one, is that the binding site on the protein
should be specific (or at least the effects of more than
one site must be separable) and yet the binding must not
be too tight, for the metal ion must be in rapid exchange
with the solvent. These stringent requirements on the
binding site are rarely met and other difficulties such as
the definition of the symmetry and magnetic symmetry axis
of the metal protein complex are relatively minor in comparison [6].

Some differences between spatial relationships, determined by the paramagnetic ion method, and X-ray diffraction are perhaps worth indicating. Everything in the NMR
method is measured relative to the metal ion which forms
a complex with the molecule under study and the complex is
tumbling in solution. In other words, spatial relationships are determined using polar rather than Cartesian coordinates. This difference, together with the rapid $(1/r^3$
or $1/r^6)$ dependence of the observed effect on distance
means that, in the presence of motion, a different 'average' conformation is observed by NMR [19]. The time scale
of the 'average' is of course also different; in NMR it is
envisaged on a millisecond time scale whereas in X-ray
methods it may be several days. Another difference between
the NMR method and X-ray methods is that in NMR the mapping is only achieved between metal ion and resolved and
assigned groups. The latter tend to be on side chains and
information about the peptide backbone is not normally
obtained.

The other method which can define conformation using
NMR [20], the measurement of three bond spin-spin coupling
constants, has not been shown to be viable in proteins.

Determination of Protein Mobility by NMR

The view of a protein derived from X-ray diffraction has
has been of a close packed, but irregular, hydrophobic
core with a hydrophilic exterior [21]. Most solution

studies confirm that the overall fold of the protein is
the same in solution as in the crystal [6]. Several methods have shown, however, that molecular flexibility must
exist in proteins. For example Linderstrøm-Lang and colleagues showed that some degree of mobility was necessary
to allow isotopic substitution of exchangeable hydrogens
[22]. The presence of flexibility in enzymes which 'flow'
through a reaction pathway [1] and in proteins which exhibit cooperative effects might be expected, but a recent
detailed X-ray analysis of the 'simple' binding protein,
myoglobin, has shown that O_2 must 'displace' some groups
before binding to the haem [23]. This is consistent with
a fluorescence study of proteins in which O_2 was shown to
penetrate readily into the core, presumably because of the
inherent flexibility of protein molecules [24].

There are essentially two ways in which NMR can be
used to detect group mobility. One depends on exchange
phenomena and the other depends on some magnetic interaction being modulated by motion, thus leading to relaxation effects.

The exchange processes may be of three distinct types:

a) *Isotope exchange* where, for example, the signal from
an exchangeable ^1H disappears as it is replaced by ^2H. The
time scale which can be observed using this signal disappearance method varies from about a second, up to the
limits of the experimenter's patience. NMR has the advantage over other methods that the half life of individual
H-bonded protons can be measured [25,26].

b) In addition to slowly exchanging hydrogens, a protein has many rapidly exchanging hydrogens on surface
groups, the rate of this *proton transfer* can often be
measured by NMR. For example, histidine C-(2) resonances
usually directly reflect the degree of protonation of the
imidazole ring because of rapid exchange between the protonated and unprotonated forms. This exchange is, however,
often slow enough to cause line broadening especially at

high fields. At the pK$_a$ value, for example, the exchange broadening is given by $\pi\Delta^2/(2k)$ where Δ is the total shift (in Hz) between the protonated and unprotonated species and k is the forward rate constant (equal to the backward rate constant at this pH). Campbell *et al.* [27,28] used this equation to measure proton transfer rates in carbonic anhydrase. It was observed that buffer increased the transfer rate and that the high activity isoenzyme, from humans, transferred protons faster than the low activity isoenzyme under the same conditions. The rates measurable by this line broadening method are in the range 10^{-1}–10^{-5} sec, depending on the size of Δ. If the exchange is in the slow, rather than the fast, limit, cross saturation is a useful technique for extending the measurable rates (see Fig. 2). (Other transfer processes such as ligand to protein can of course also be analysed by line broadening and cross saturation.)

c) The third type of exchange process discussed here is where there is no transfer from protein to solvent, but there is some *change of environment* of a group in the protein, so that at least two different secondary shifts are experienced. As far as the NMR experiment is concerned the methods of analysis for this exchange process are the same as for proton transfer, although, unless the conformational change is induced, no titrations can be performed. A good example of this sort of exchange process is provided by a tyrosine residue in horse cytochrome *c*. This tyrosine undergoes a temperature dependent rotational flip about its C$_\beta$–C$_\gamma$ bond (see Fig. 2). This motion causes severe line broadening at some temperatures and analyses of the line broadening and cross saturation experiments have been carried out to obtain the data shown in Fig. 2 [29]. This ring flipping would not be expected from observation of the X-ray structure since the group is close packed. A theoretical analysis of this sort of molecular motion in the bovine pancreatic trypsin inhibitor has been

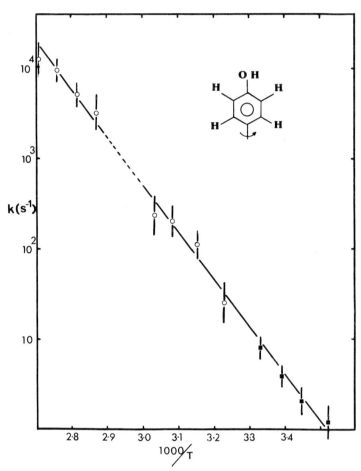

Fig. 2. The temperature dependence of the flip rate of tyrosine 46 in horse ferrocytochrome *c* as determined by lineshape analysis (o) and cross saturation experiments (■). The data are from [28].

given by Gelin and Karplus [30] and Wüthrich, and colleagues [8]. Another example of this type of exchange process is the induced EI ⇌ E*I transition observed on binding inhibitor to lysozyme [6].

Molecular motion modulates local magnetic interactions.
For example, if two spins are a distance r apart, tumbling
of the internuclear vector **r** leads to a fluctuation of the
magnetic field experienced by each of the nuclei. This
leads to a finite probability of transitions taking place

between the energy levels of the system. Measurement of
the parameters T_1, T_2 and η (the nuclear Overhauser enhancement) leads indirectly to a definition of the molecular motion. This step usually depends on some assumptions, including the distance r and the number of nuclei
(N) causing the relaxation of the observed nucleus. The
analysis is most straightforward for ^{13}C studies of carbons with directly bonded protons since r and N are then
well defined. Such studies [3,31,32] suggest that α carbons and aromatic groups tumble essentially isotropically,
with the same correlation time as the overall tumbling of
the protein in solution, but that surface groups such as
lysine have much more extensive motion. This result is
supported by ^{15}N studies [33] and by ^{19}F studies [34] (in
the ^{19}F studies chemical shift anisotropy is a significant relaxation mechanism at high field whereas in all
other cases discussed here dipolar mechanisms dominate).

An interesting case of relaxation studies is ^{1}H NMR
where r and N are usually not well defined. Measurement
of the relaxation rates as a function of frequency has
been shown to help in overcoming these problems [35]. A
general observation is that many of the ^{1}H resonances have
similar T_1 values, which are almost independent of temperature and the molecular weight of the protein. These results can only be explained by some anisotropic motion
within the protein structure at a frequency around 10^9
s^{-1}. Unfortunately it is not straightforward to locate the
position of this anisotropic motion since, as pointed out
by Kalk and Berendsen [35], cross relaxation [37] may be
very strong in ^{1}H NMR of proteins. It is unclear, at this
time, whether all the anisotropic motion can be accounted
for by methyl rotation [37] or whether there is some sort
of general 'wobble' taking place in the hydrophobic core
of the molecule at 10^9 s^{-1}. The latter possibility is not
necessarily ruled out by the ^{13}C results since ^{1}H NMR is
quite likely to be sensitive to different types of mole-

cular motion.

Determination of Thermodynamic Parameters by NMR

One of the main advantages of spectroscopic methods is that the effect of environmental changes are readily monitored. Thus, titrations of pH or ligand concentration can be used to give pK_a values of individual groups or ligand dissociation constants, while variation of temperature or ionic strength yields information about the stability of the structure in solution. This sort of information is important if we are to understand how the protein structure modifies pK_a values, redox potentials, ligand affinities, etc. In addition, since NMR can observe all the atoms and since the resonance positions depend on the local electronic environment, it should be possible to detect 'unusual' environments in the structure. It is difficult to make very general statements about these thermodynamic aspects and they will be illustrated very briefly here using three specific examples.

Haemoglobin takes up protons on releasing oxygen. This phenomenon is known as the Bohr effect and Perutz in his classic X-ray studies [38] predicted that histidine β146 might be involved, due to a change in local environment of this group between the oxy and deoxy states of the protein structure. The pK_a of this group can be measured using NMR [39,40] by following the resonance of the C-(2) proton of histidine β146 in both oxy and deoxy states (see Fig. 3). This is a good example of the complementary nature of NMR and X-ray crystallography.

Triose phosphate isomerase catalyses the interconversion of dihydroxy-acetone phosphate (DHAP) and glyceraldehyde phosphate. This catalysed reaction has been the subject of intensive kinetic investigations by Knowles and colleagues [41]. The reaction proceeds via an ene-diol intermediate and the inhibitor 2-phosphoglycollate has been proposed as a transition state analogue [42]. The

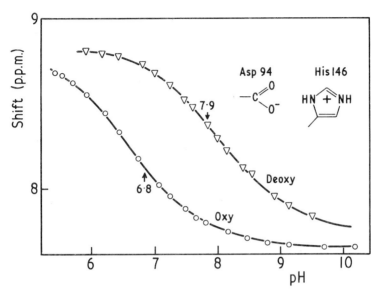

Fig. 3. The titration of histidine β146 in haemoglobin in the oxy and deoxy form, as observed by the shift of the C-(2) proton resonance. The assignment is from the work of Kilmartin *et al.* [38] and the data are from [39]. The proximity of Asp 94 raises the pK_a of the deoxy form.

enzyme has also been studied using X-ray crystallography and a 2.5 Å resolution electron density map has been used to construct a model of the native dimer of two 25,000 dalton subunits [43]. X-ray studies of the enzyme in the presence of ligands have also been carried out to 6 Å resolution by D.C. Phillips and colleagues, but at present there is still a considerable gap in our knowledge, which needs to be bridged before the structural and kinetic information can be completely understood. In a recent ^{31}P study of DHAP and 2-phosphoglycollate binding to the enzyme [44] it has been shown that the mono and dianion of the phosphate of DHAP bind equally well to the enzyme while the phosphate of phosphoglycollate binds as the dianion. This result was unexpected and any model of enzyme substrate complexes which emerges from an amalgam of the various techniques must explain this piece of information.

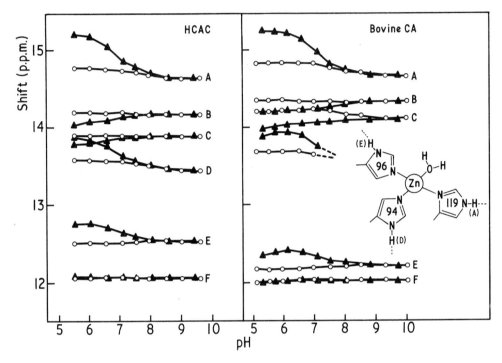

Fig. 4. The pH dependence of 6 exchangeable resonances in bovine and human carbonic anhydrase C(▲) observed at 270 MHz using a pre-saturation pulse method to suppress the solvent peak. The resonances A, B, C, D and E are sensitive to pH and are influenced by the presence of 10 mM iodide as shown (o). The insert shows the tentative assignments of resonances A, D and E to the zinc liganded histidines, 119, 94 and 96.

Carbonic anhydrase is a zinc containing enzyme (30,000 daltons) which catalyses the interconversion of CO_2 and HCO_3^-. A recent NMR study of the exchangeable ^1H resonances [45] shows that there are several such resonances in the 12-15 ppm range (see Fig. 4), which are sensitive to the ionization of a group which is essential for activity ($pK_a \simeq 7$). Resonances A, D and E were assigned to the NH protons on the liganded imidazole rings as shown in Fig. 4. These results then yield three interesting pieces of information: the structures of the enzyme from the cow and the high activity human isoenzyme must be very similar

around the zinc atom; the ionization of one of the imidazole ligands around pH 7, as proposed by Appleton and Sarkar
[45], is rather unlikely; the NH resonance of histidine
119 occurs at a very low field indeed (as do resonances B
and C, which are also near the zinc). This low field position suggests that this proton is involved in a strong
hydrogen bond and it is reminiscent of the H-bonded resonance observed by Robillard and Shulman [47], at very low
field, in the spectrum of several serine proteases. It is
not impossible that these resonances are reflecting some
sort of local high energy state, which may be essential
for catalysis [48].

Conclusions

It is only in the last few years that NMR techniques
have been powerful enough to study satisfactorily protein
structure at the atomic level. It has been demonstrated
that the method is now able to give information about
spatial relationships, relative mobility and local energetic effects. NMR will, no doubt, become increasingly
used as a tool, truly *complementary* to other powerful
methods for investigating the structure and function of
proteins, such as amino acid sequencing, X-ray crystallography and enzyme kinetics.

Acknowledgements

I have drawn freely from work done in collaboration
with F.F. Brown, C.M. Dobson, P.A. Kiener, S. Lindskog,
G.R. Moore, S.G. Waley and R.J.P. Williams. Geoff Moore
kindly gave me the data shown in Fig. 1 and Table 1. It
should be clear from the reference list that I have worked
especially closely with Chris Dobson and Bob Williams in
the last few years and I express my gratitude to them, and
all my Oxford colleagues, for their example. This is a
contribution from the Oxford Enzyme Group, which is supported by the Science Research Council.

References

1. Campbell, I.D., Dobson, C.M. and Williams, R.J.P. (1977). *In: Proceedings of the Solvay Conference*. Academic Press (in press).
2. Campbell, I.D., Dobson, C.M. and Williams, R.J.P. (1975). *Proc. R. Soc. Lond.* **A 345**, 23–40.
3. Oldfield, E., Norton, R.S. and Allerhand, A. (1975). *J. Biol. Chem.* **250**, 6368–402.
4. Shindo, H. and Cohen, J.S. (1976). *Proc. Nat. Acad. Sci. U.S.A.* **73**, 1979–83.
5. Campbell, I.D. and Dobson, C.M. (1977). *In: Methods of Biochemical Analysis* (D. Glick, ed.), Academic Press (in press).
6. Dobson, C.M. (this volume).
7. Roberts, G.C.K., Feeney, J., Birdsall, B., Kimber, B., Griffiths, D.V., King, R.W. and Burgen, A.S.V. (this volume).
8. Wüthrich, K., Wagner, G., Richarz, R. and de Marco, A. (this volume).
9. Moore, G.R. and Williams, R.J.P. (1975). *FEBS Lett.* **53**, 334–8.
10. Gettins, P. and Dwek, R.A. (this volume).
11. Robillard, G.T. (this volume).
12. Ambler, R.P., Bruschi, M., Campbell, I.D., Cookson, D.J., Le Gall, J., Moore, G.R., Pitt, R.C. and Williams, R.J.P. (to be published).
13. Kellar, R.M. and Wüthrich, K. (private communication).
14. Markley, J.L. (1975). *Biochemistry* **14**, 3546–54.
15. Moore, G.R. and Williams, R.J.P. (1977). *In: Proceedings of the Marseille Conference of Electron Transfer Systems*. C.N.R.S., Paris, in press.
16. Dickerson, R.E. and Timkovich, R. (1975). *In: The Enzymes* (P. Boyer, ed.), 3rd edn. Vol. XI, Academic Press, 397–549.
17. Levine, B.A. and Williams, R.J.P. (1975). *Proc. R. Soc. Lond.* **A 345**, 5–22.
18. Dobson, C.M. and Levine, B.A. (1976). *In: New Techniques in Biophysics and Cell Biology* (Paine, R.H. and Smith, B.J., eds). Wiley, London, **3**, 19–91.
19. Williams, R.J.P. (1976). Merck, Sharpe & Dohme Research Lecture.
20. Feeney, J. (1975). *Proc. R. Soc. Lond.* **A 345**, 61–72.
21. Chothia, C. (1975). *Nature* **254**, 304–8.
22. See, for example, Hvidt, A. & Nielsen, S.O. (1966). *Adv. in Protein Chem.* **21**, 287–386.
23. Takano, T. (1977). *J. Mol. Biol.* **110**, 537–84.
24. Campbell, I.D., Dobson, C.M. and Williams, R.J.P. (1975). *Proc. R. Soc. Lond.* **B 189**, 485–502.
25. Lascowicz, J.R. and Weber, G. (1973). *Biochemistry* **12**, 4161–79.
26. Patel, D.J. and Canuel, L.L. (1976). *Proc. Nat. Acad. Sc. U.S.A.* **73**, 1398–1402.
27. Campbell, I.D., Lindskog, S. and White, A.I. (1974). *J. Mol. Biol.* **90**, 469–89.
28. Campbell, I.D., Lindskog, S. and White, A.I. (1975). *J. Mol. Biol.* **98**, 597–614.
29. Campbell, I.D., Dobson, C.M., Moore, G.R., Perkins, S.J. and Williams, R.J.P. (1976). *FEBS Lett.* **70**, 96–100.
30. Gelin, B.R. and Karplus, M. (1975). *Proc. Nat. Acad. Sci. U.S.A.* **72**, 2002–6.

31. Cozzone, P.J., Opella, S.J., Jardetzky, O., Berthou, J. and Jolles, P. (1975). *Proc. Nat. Acad. Sci. U.S.A.* **72**, 2095–8.
32. Visscher, R.B. and Gurd, F.R.N. (1975). *J. Biol. Chem.* **250**, 2238–42.
33. Gust, D., Moon, R.B. and Roberts, J.D. (1975). *Proc. Nat. Acad. Sci. U.S.A.* **72**, 4696–700.
34. Hull, W.E. and Sykes, B.D. (1975). *J. Mol. Biol.* **98**, 121–53.
35. Coates, H.B., McLaughlin, K.A., Campbell, I.D. and McColl, C.E. (1973). *Biochem. Biophys. Acta.* **310**, 1–10.
36. Kalk, A. and Berendsen, H.J.C. (1976). *J. Mag. Res.* **24**, 343–66.
37. Campbell, I.D. and Freeman, R. (1973). *J. Mag. Res.* **11**, 143–62.
38. Perutz, M.F. (1970). *Nature* **228**, 726–39.
39. Kilmartin, J.V., Breen, J.J., Roberts, G.C.K. and Ho, C. (1975). *Proc. Nat. Acad. Sci. U.S.A.* **70**, 1246–9.
40. Brown, F.F. and Campbell, I.D. (1976). *FEBS Lett.* **65**, 322–6.
41. Alberry, W.J. and Knowles, J.R. (1976). *Biochemistry* **15**, 5627–40.
42. Wolfenden, R. (1970). *Biochemistry* **9**, 3404–7.
43. Banner, D.W., Bloomer, A.C., Petsko, G.A., Phillips, D.C., Pogson, C.I. and Wilson, I.A. (1975). *Nature* **255**, 609–14.
44. Campbell, I.D., Kiener, P.A. and Waley, S.G. (1977). *In: Proceedings of the Biochem. Soc.*, Aberdeen meeting, March 1977, in press.
45. Campbell, I.D., Lindskog, S. and White, A.I. (1977). *Biochem. Biophys. Acta.*, in press.
46. Appleton, D.W. and Sarkar, B. (1974). *Proc. Nat. Acad. Sci., U.S.A.* **71**, 1686–90.
47. Robillard, G. and Shulman, R.G. (1974). *J. Mol. Biol.* **86**, 519–40.
48. Vallee, B.L. and Williams, R.J.P. (1968). *Proc. Nat. Acad. Sci. U.S.A.* **59**, 498–500.

COMPLETION OF X-RAY STRUCTURES OF PROTEINS BY HIGH RESOLUTION NMR

KURT WÜTHRICH, GERHARD WAGNER, RENÉ RICHARZ
and ANTONIO DE MARCO

*Institut für Molekularbiologie und Biophysik
Eidgenössische Technische Hochschule
8093 Zürich, Switzerland*

Introduction

In the course of the last decade high resolution NMR
techniques have become a standard method for investigations of protein conformation [1-3]. It is one of the
attractive features of NMR that proteins can be studied
in solution, or more generally under conditions which may
be closely related to the physiological environment.
Thus structural information is obtained on proteins which
are for want of suitable single crystals or other technical reasons not amenable to X-ray studies. On the other
hand, with the spectral resolution obtained with modern
high field spectrometers, NMR can provide essential complementary structural information on proteins for which
X-ray data are also available. The combined use of X-ray
and NMR techniques is thus a particularly promising approach to obtain detailed insight into a variety of aspects
of protein conformation. In addition, investigations of
molecules for which refined X-ray atomic coordinates are
available have an important role in view of further developments of NMR methodology for studies of proteins.

Structural information from NMR studies may include
comparison of the protein conformations in crystals and
in solution, studies of medium dependent conformational
transitions, molecular dynamics, mechanistic aspects of

protein folding, and intermolecular interactions of proteins. Much of our efforts over the last few years have
been concentrated on the basic pancreatic trypsin inhibitor (BPTI), a small globular protein with many characteristics which make it a suitable model case for these various aspects of NMR applications. Atomic coordinates for
BPTI were obtained from highly refined X-ray data [4], the
globular conformation of BPTI in solution is maintained
over a large range of pH, < 1 to ≳ 12 at 25°, and temperature, 1° to > 90° at neutral pH, [5,6] and a selection of
well characterized chemically modified BPTI species are
available [5,7]. Also, the stable complexes formed with
proteases [8] appear attractive for studies of protein—
protein interactions, where the NMR properties of isolated BPTI can in principle be used as natural spectroscopic probes [9].

Whether or not the NMR data are analysed in combination
with X-ray atomic coordinates, much depends on reliable
identification of the spin systems of particular amino
acid residues, and possibly resonance assignments to specific positions in the amino acid sequence. The potential
for observation and assignment of numerous individual
resonance lines in the NMR spectra of macromolecules has
been greatly improved by the newest generation of high
field spectrometers, which really provides the basis for
obtaining essential structural information to complete
the crystal data from X-ray studies. In the following main
part of this paper some aspects of resonance assignments
are illustrated with studies of BPTI at a field strength
of 8.4 tesla.

Assignment of the Methyl Resonances in the ^{1}H and ^{13}C NMR Spectra of BPTI

The twenty common amino acid residues contain characteristic spin systems, most of which can readily be identified on the basis of qualitative NMR spectral features

and the pH dependence of certain parameters [3]. If the
individual resonance lines are resolved in the spectrum
of the proteins, the spin systems of different types of
amino acid residues can hence be identified.

Various digital filtering techniques have been proposed
to improve the spectral resolution [10-12]. As an illustration, Fig. 1 shows the effect of filtering with the
sine bell routine [12] on the ^{1}H NMR spectrum of BPTI.

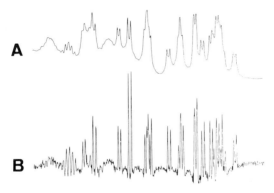

Fig. 1. Aliphatic region from 0.5 to 2.0 ppm of the FT ^{1}H NMR spectrum at 360 MHz of the basic pancreatic trypsin inhibitor, 0.01 M
solution in ^2H_2O, 225 scans, 4000 Hz spectral width, 2 sec. acquisition time, 16 K of memory in the time domain. A: Normal spectrum.
The natural line width $\Delta\omega_{\frac{1}{2}}$ is approximately 20 rad sec^{-1}. B: Same
spectrum as A after digital filtering of the FID with the sine bell
routine [12].

The entire ^{1}H NMR spectrum of a solution of BPTI in
^2H_2O is shown in Fig. 2A [13]. The region from 0 to 6.0
ppm contains the resonances of all the C_α-protons and the
non-aromatic amino acid side chains. The residual water
protons give rise to the sharp strong line at 4.6 ppm.
The region from 6.0 to 7.8 ppm contains the resonances of
the aromatic side chains, and between 7.0 and 11.0 ppm
tnere are tne resonances of hydrogen bonded amide protons
of the α-helix and the β-sheet in BPTI [6,14]. These exchange slowly with deuterium of ^2H_2O, as is illustrated
in the spectra 2B and 2C, which were recorded at different
times after the protein had been dissolved in ^2H_2O. The

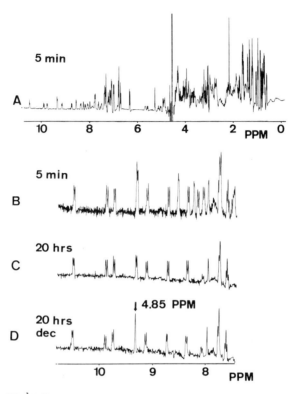

Fig. 2. A. FT ^{1}H NMR spectrum at 360 MHz of a 0.005 M solution of
BPTI in ^{2}H$_2$O, p^{2}H = 4.5, T = 45°, after resolution enhancement. The
spectrum was recorded 5 minutes after the protein had been dissolved
in ^{2}H$_2$O. B. Expanded representation of the region from 7.6 to 11.0
ppm in spectrum A, which contains the resonances of hydrogen bonded
amide protons of the α-helix and the β-sheet in BPTI. C. Same as B,
recorded 20 hrs after the protein had been dissolved in ^{2}H$_2$O. D. Same
as C, except that a double resonance irradiation field was applied to
the C$_\alpha$-proton resonance at 4.85 ppm. It is seen that the amide proton resonance at 9.4 ppm was thus decoupled.

expanded spectra 2B and 2C show that the fine structure
of the individual amide proton resonances is well resolved. Different resolved multiplets which are part of the
same spin system can routinely be related by double resonance experiments [1–3]. This is illustrated in Fig. 2D,
which shows that the amide proton line at 9.4 ppm and the
α-proton line at 4.85 ppm originate from the same amino
acid residue.

In the following attention is focused on the aliphatic
region between 0 and 4.5 ppm (Fig. 2A), where spin decoupling was used to assign individual ones of the total
of 20 methyl resonances in BPTI to different types of.
amino acid residues [13]. Amino acids in BPTI which contain methyl groups are 6 Ala, 1 Val, 2 Leu, 2 Ile, 3 Thr
and 1 Met. Even though the methyl lines were well resolved
after digital filtering (Fig. 1B), the methine and methylene proton resonances showed extensive mutual overlap
(Fig. 2A). Therefore, double resonance difference spectroscopy [15,16] was used to identify the different spin systems which contain methyl groups (Fig. 3). In the spectrum
of Fig. 3B, which corresponds to the difference between
the decoupled and the off-resonance decoupled spectrum,
it is seen that three methyl resonances were simultaneously decoupled by irradiation at 4.02 ppm. The difference
multiplet patterns expected under conditions of complete
decoupling were simulated in Fig. 3B. It is seen that for
each methyl group one has a negative doublet for the un-
decoupled resonance and a positive singlet for the decoupled resonance. These features are also readily recognized in the experimental spectrum of Fig. 3B. However,
since very low power was used in the decoupling channel
to obtain the desired selective irradiation of individual
resonances in crowded spectral regions, incomplete decoupling resulted in the appearance of additional weak
lines symmetrical with respect to the principal resonance
in the decoupled spectrum [17]. Corresponding 'wings' are
also observed in the spectra C to E of Fig. 3. The difference multiplets at 4.01 and 3.99 ppm in Fig. 3, C and D,
respectively, correspond otherwise to what one expects for
the A_3X spin system (for the notation used see e.g. [3]
or [17]) of Ala (Fig. 3, C' and D'). The multiplet at 4.04
ppm in Fig. 3E corresponds to the A_3MX spin system of Thr
(Fig. 3E'), where the coupling constants $^3J_{AM}$ and $^3J_{MX}$
appear to be essentially identical. From double resonance

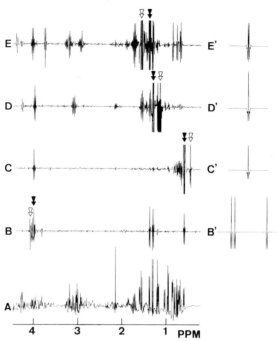

Fig. 3. Assignment of three methyl resonances in BPTI by double resonance difference spectroscopy. A. Region from 0.5 to 4.5 ppm in the
FT ^{1}H NMR spectrum at 360 MHz of a 0.005 M solution of BPTI in ^{2}H$_2$O
in which all the amide protons had been exchanged with deuterium,
p^{2}H = 7.0, T = 45°, after resolution enhancement. B-E. Difference
spectra obtained by subtracting from a spectrum recorded with double
resonance irradiation at the position of the filled arrow a spectrum
obtained with off-resonance irradiation at the position of the empty
arrow. The difference patterns of interest, which are to be compared
with the simulated difference multiplets B'-E' (see text), have been
shadowed. Additional multiplets appear in the spectra D and E, since
the double resonance irradiation had to be applied in a very crowded
region of the spectrum.

experiments of the type of Fig. 3 and considerations of
the qualitative spectral features of the amino acid residues [3], all the 20 methyl proton resonances were
assigned to specific types of amino acids (Fig. 4).

The spectral resolution is also markedly improved in
the ^{13}C NMR spectra at higher fields, in particular for
the protonated carbon positions (Fig. 5). The entire ^{13}C
NMR spectrum of BPTI at 90.5 MHz is shown in Fig. 5, and

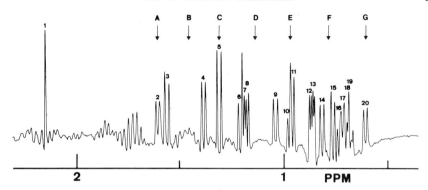

Fig. 4. Spectral region from 0 to 2.5 ppm of the FT ^{1}H NMR spectrum at 360 MHz of BPTI in ^{2}H$_2$O, p H = 6.5, T = 35° after resolution enhancement. The numbers indicate the resonances of the 20 methyl groups in the protein, which were assigned to different types of amino acids as follows: 2,3,5,6,9 and 20, Ala; 10 to 19, Val, Leu and Ile; 4, 7 and 8, Thr; 1, Met.

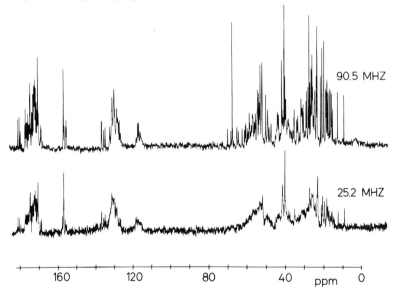

Fig. 5. Natural abundance ^{1}H noise-decoupled FT ^{13}C NMR spectra at 25.2 MHz and 90.5 MHz of a 0.025-M solution of the basic pancreatic trypsin inhibitor (BPTI) in ^{2}H$_2$O, p^{2}H = 8.2, T = 35°, accumulation time 12 hrs.

the region from 5 to 25 ppm, which contains the methyl carbon resonances, is represented on an expanded scale in Fig. 6. On the basis of the spectral resolution attained in both the ^{1}H (Fig. 4) and ^{13}C (Fig. 6) spectra, corres-

KURT WÜTHRICH ET AL.

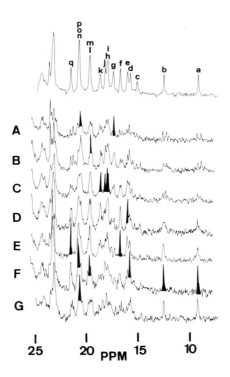

Fig. 6. Top trace: Spectral region from 0 to 25 ppm of the ^{1}H noise-decoupled FT ^{13}C NMR spectrum at 90.6 MHz of a 0.025-M solution of BPTI in ^{2}H$_2$O, p^{2}H = 6.5, T = 35°. The letters indicate the positions of individual methyl carbon resonances. A-G: Same as the top trace with selective ^{1}H irradiation of 0.8 watt at A. 1.619 ppm, B. 1.464 ppm, C. 1.311 ppm, D. 1.133 ppm, E. 0.958 ppm, F. 0.758 ppm. G. 0.597 ppm (see Fig. 4). The collapsed multiplets in the individual spectra are shadowed.

ponding ^{1}H and ^{13}C resonances of methyl groups could be identified by heteronuclear double resonance experiments [18]. The analysis of these data used that in the ^{13}C NMR spectra obtained with selective ^{1}H off-resonance irradiation, the methyl resonances can be recognized from their quartet fine structure [3] and over a certain range the residual ^{1}H-^{13}C spin–spin coupling varies linearly with the proton irradiation frequency [19,20]. Fig. 6 shows the ^{13}C NMR spectra obtained with selective proton irradiation at the positions indicated by the arrows in

Fig. 4. From inspection of Fig. 6 the ten well resolved
one-carbon lines a to g, j, k and q could from their
multiplet structures be unambiguously identified as
methyl carbons. These resonances could then by least
squares fits of the ^{1}H frequency dependence of the residual spin—spin couplings be related to individual [1]H
resonances (figure caption 7). Between 16 and 22 ppm

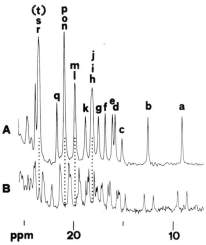

Fig. 7. Spectral region from 0 to 25 ppm of the FT ^{13}C NMR spectra at
90.5 MHz of a 0.05-M solution of BPTI in ^2H_2O, $p^2H = 4.2$, T = 35°.
A. [1]H noise-decoupled spectrum. Because of the lower p^2H, resonance
j is at somewhat higher field than in the corresponding spectrum of
Fig. 6 and overlaps with the resonances i and h. B. Off-resonance
selective ^{1}H irradiation of 1.0 watt at 4.0 ppm. The letters indicate
the individual methyl carbon lines, which were assigned as follows
(in parenthesis: corresponding ^{1}H NMR line, see Fig. 4): e(9),
g(2 or 3), j(5), k(6), n(2 or 3) and o(20), Ala; a and b (18 and 19),
Ile-δ; h(7 or 8), i(7 or 8) and m(4), Thr; c(1), Met; d, f, l, p to
t, Val, Leu and Ile-γ^2.

there are three groups of overlapping resonance lines,
i.e. h and i, 1 and m, and n to p. The interpretation of
this more crowded region was confirmed by spectral simulations and the additional ^{1}H off-resonance decoupling
experiment in Fig. 7, which showed that under conditions
of incomplete decoupling the resonances h, i, 1 to s, and
probably also t, give rise to quartets and therefore must
correspond to methyl groups in the protein. The assign-

ments of the methyl carbon resonances are indicated in
the figure caption 7.

Discussion

The experiments described here dealt exclusively with
the assignments of individual resonance lines to spin
systems of the different types of amino acid residues.
Provided that the individual components of the spin sys-
tens are well resolved, resonance identifications on this
level can usually be obtained solely with spectroscopic
techniques, as outlined in this paper. In BPTI, the spin
systems of the aromatic rings [21,22] and in part of the
Lys residues [23] were thus also identified.

The next level of resonance assignments would be to
relate individual spin systems with specific locations
in the amino acid sequence. Except in particularly favourable situations (e.g. in BPTI, the single Met resonance could readily be assigned to position 52, and the
resonances of the C-terminal Ala 58, i.e. 5 and j in
Figs. 4 and 7, were assigned from their pH titration
shifts) this second level of assignment can only be
achieved by studies of chemically modified proteins or,
if available, suitable sets of homologous proteins [1-3].
In BPTI, resonance assignments from studies of modified
proteins include those of Ala 16 in the active site
(resonances 6 and k in Figs. 4 and 7) [13,18] and the
four Tyr residues [24].

Structural information derived from the NMR data on
BPTI so far includes direct evidence that the backbone
conformations in the crystal structure and in solution
are closely similar [6,14,24], demonstration of local
differences between the protein surface structure in the
crystal and in solution [23] and quantitative descriptions of the molecular dynamics in solution [21,22]. For
the protein NMR spectroscopist, the BPTI studies so far
revealed a wealth of data on conformation dependent chemi-

cal shifts. e.g. conformation dependent chemical shifts
of corresponding ^{1}H and ^{13}C methyl resonances (Fig. 7),
which may contribute to establish novel empirical rules
relating NMR features with aspects of protein conformation.

Acknowledgement

Financial support by the Schweizerischer Nationalfonds
(project 3.1510.73) is gratefully acknowledged.

References

1. Dwek, R.A. (1973). *Nuclear Magnetic Resonance in Biochemistry.
   Applications to Enzyme Systems.* Clarendon Press, Oxford.
2. James, T.L. (1975). *Nuclear Magnetic Resonance in Biochemistry.*
   Academic Press, New York.
3. Wüthrich, K. (1976). *NMR in Biological Research: Peptides and
   Proteins.* North Holland, Amsterdam.
4. Deisenhofer, J., Steigemann, W. (1975). *Acta Cryst.* B **31**, 238–50.
5. Vincent, J.P., Chicheportiche, R., Lazdunski, M. (1971). *Eur. J.
   Biochem.* **23**, 401–11.
6. Masson, A., Wüthrich, K. (1973). *FEBS Lett.* **31**, 114–18.
7. Jering, H., Tschesche, H. (1976). *Eur. J. Biochem.* **61**, 443–52.
8. Tschesche, H. (1974). *Angew. Chem.* **13**, 10–28.
9. De Marco, A., Wüthrich, K. (1976). *In: Abstracts of the VIIth
   Internat. Conference on Magnetic Resonance in Biological Systems, St. Jovite, Quebec, Canada,* p. TH–R6.
10. Ernst, R.R. (1966). *In: Advances in Magnetic Resonance,* Vol. 2
    (ed. J.S. Waugh), pp. 1–135. Academic Press, New York.
11. Campbell, I.D., Dobson, C.M., Williams, R.J.P., Xavier, A.V.
    (1973). *J. Magn. Reson.* **11**, 172–81.
12. De Marco, A., Wüthrich, K. (1976). *J. Magn. Reson.* **24**, 201–4.
13. De Marco, A., Tschesche, H., Wagner, G., Wüthrich, K. (1971).
    Biophysics of Struct. Mech. (in press).
14. Karplus, S., Snyder, G.H., Sykes, B.C. (1973). *Biochemistry* **12**,
    1323–9.
15. Gibbons, W.A., Beyer, C.F., Dadok, J., Sprecher, R.F., Wyssbrod,
    H.R. (1975). *Biochemistry* **14**, 420–9.
16. Campbell, I.D., Dobson, C.M. and Williams, R.J.P. (1975). *Proc.
    Roy. Soc. Lond.* **A.345**, 23–40.
17. Pople, J.A., Schneider, W.G., Bernstein, H.J. (1959). *High Reso-
    lution Nuclear Magnetic Resonance.* McGraw-Hill, New York, p. 229.
18. Richarz, R., Wüthrich, K. *Proc. Nat. Acad. Sci.* (submitted).
19. Ernst, R.R. (1966). *J. Chem. Phys.* **45**, 3845–61.
20. Birdsall, B., Birdsall, N.J.M., Feeney, J. (1972). *J.C.S. Chem.
    Comm.* 316–17.
21. Wagner, G., De Marco, A., Wüthrich, K. (1975). *J. Magn. Reson.*
    20, 565–9.
22. Wagner, G., De Marco, A., Wüthrich, K. (1976). *Biophys. Struct.*

Mechanism **2**, 139–58.
23. Brown, L.R., De Marco, A., Wagner, G., Wüthrich, K. (1976). *Eur. J. Biochem.* **62**, 103–7.
24. Snyder, G.H., Rowan, R., Karplus, S., Sykes, B.D. (1976). *Biochemistry* **14**, 3765–77.

THE STRUCTURE OF LYSOZYME IN SOLUTION

C.M. DOBSON

The Inorganic Chemistry Laboratory, Oxford

Introduction

The methods of X-ray crystallography and nuclear magnetic resonance (NMR) are both used to provide information
about protein structure and structural changes. The two
techniques are very different in their approaches and in
their information content. In this paper the results of
proton NMR studies of one enzyme, lysozyme from hen egg
white, will be discussed and compared to the results obtained from X-ray crystallographic studies. This involves
a comparison of the structure and behaviour of the enzyme
in the crystalline phase and in solution. Such comparisons should allow conclusions to be made not only for this
enzyme but also for proteins in general.

Several aspects of protein structure will be considered. The first aspect is concerned with the 'average'
positions of atoms in space in the native protein. X-ray
crystallography provides a single set of coordinates for
each atom obtained by fitting the primary structure (sequence) to the observed electron density map. The electron density map is obtained over a period of time from a
crystal containing a large number of molecules. The electron density will be 'averaged' with respect to time, if
molecular motion occurs in the crystal, and averaged with
respect to space, if different molecules exist in differ-

ent conformations. NMR generally observes single resonances from each nucleus in the molecule, and averaging will also be a feature of this technique, although the timescale is very much shorter. In the comparison of the 'averaged structures' seen by each technique allowance will only be made for certain motions at this stage. Thus, rotation of methyl groups will be permitted, as will flipping of aromatic rings, for there is direct experimental evidence that these motions occur even on the timescale of the NMR experiment (see below).

The second aspect of structure is concerned with the motion of individual atoms in the structure. Comparison of the evidence for motion which is available from the two techniques will be made.

The third aspect of structure is concerned with conformational changes which can be induced by known perturbations such as change of pH, binding of inhibitors or chemical modification. The groups involved in a conformational change can be compared in crystals and in solution. In addition NMR can be used to determine rates of the conformational changes, information which cannot be obtained from crystallographic studies. However such information is available in certain cases from other techniques, such as temperature jump studies, and this will be compared with the NMR conclusions.

Resolution and Assignment

Before any detailed information can be obtained from the NMR spectrum, it is essential to resolve resonances of individual nuclei and to assign these resonances to specific groups in the protein molecule.

The resolution problem results because a protein spectrum contains a very large number of proton resonances with relatively large linewidths which fall into a rather narrow band of frequencies. There is therefore considerable overlap of resonances. Solution of this problem re-

quires a reduction of the number of resonances, a reduction of linewidths or an increase in the frequency (chemical shift) range. Each of these remedies is successful to
a certain degree [1]. These have been discussed elsewhere,
and are summarised in Table 1. The application of these
methods will be clear from spectra presented in this work.

TABLE 1

Methods for resolution of protein spectra

Method	Example	Reference
Reduction of linewidths	Convolution difference	[2]
Reduction of number of resonances (sub-spectra)	Many types of difference spectroscopy	[1,2]
	Multiplet selection	[4,5]
	Spin-echo sequences	[4]
	Isotopic exchange	[31]
Separation of overlapping resonances	Addition of shift probes	[1,7,32]

Increase in resolution of the spectrum in itself aids
assignment of resonances, by making it possible to extract
the parameters (e.g. chemical shift, area, coupling constant and multiplet structure) which are required for the
first stage of assignment. The essential features of the
assignment process are shown in Table 2. It is convenient
to divide the assignment process into two stages [1].
Stage 1 is concerned with the assignment of an observed
resonance to a specific *type* of amino acid residue and
this can be subdivided into two steps. First, a resonance
is assigned to a specific type of proton, such as a methyl
group, an aromatic proton, or an indole NH proton. These
groups are the most readily studied types of proton and
can be identified with certainty from their chemical
shifts, areas and coupling patterns. Secondly, the type

of amino acid can be identified by careful consideration
of these parameters. The methods have been fully described
elsewhere, and specific examples only will be discussed
here. The aromatic protons of lysozyme can be clearly
observed between 6.0 and 9.0 ppm (Fig. 1). The only singlet resonances in this part of the spectrum arise from the
C(2)H and the C(4)H protons of histidine and from the
C(2)H protons of tryptophan. In the convolution difference
spectrum [2] of lysozyme, it is possible to observe several of the singlet resonances, but the severe overlap
problem makes it impossible to observe all the resonances.
The use of lanthanide probes (see for example [3]) enables other resonances to be seen by shifting them out of
the spectral envelope. However, by using spin echo pulse
sequences it is possible to select resonances for observation on the basis of their multiplet structure [4].
Spectra containing only singlet and triplet resonances, or
containing only doublet resonances can be obtained for
this aromatic region of the spectrum where the coupling
constants are similar for each multiplet (J ≈ 8.5 Hz, the
smaller couplings J ≤ 3 Hz are not observed). In Fig. 2
the identification of all singlet resonances is shown.
Two of these peaks titrate with pH in a manner which shows
that they arise from the single histidine residue (His
15). The other six arise from the tryptophans. Two are
overlapping (C5 and C6) but can be separated by binding
of lanthanides [5]. One, C1, is broad and will be discussed below. In the aromatic region, double resonance experiments permit coupling patterns to be determined, en-
abling observed resonances to be assigned to specific
types of residue. In particular all the resonances of the
three tyrosines have been identified [6], and appear as
two proton intensity doublets (Fig. 2). Additionally, all
six tryptophan NH resonances have been assigned, and more
than twenty of the sixty methyl groups [7].

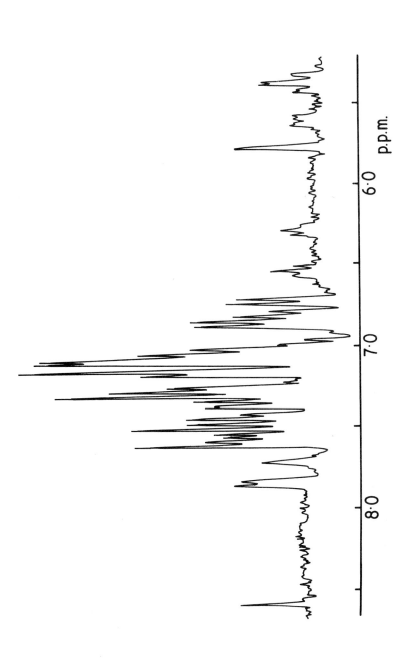

Fig. 1. Convolution difference spectrum of lysozyme in the aromatic proton region. 5mM lysozyme, pH 4.8, 66°C.

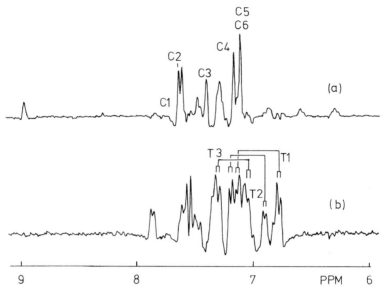

Fig. 2. Multiplet selection spectra of aromatic proton region of lysozyme. (A) singlet and triplet resonances; resonances C1–C6 arise from the C(2)H protons of the six tryptophans. (B) Doublet resonances; resonances T1–T3 arise from the three tyrosines. 5 mM lysozyme, pH 3.9, 54°C. Pulse sequences as described in [4].

Stage 2 of the assignment procedure involves assignment of resonances to specific residues in the sequence. The methods for carrying this out are given in Table 2. Note that, as with Stage 1, some of these methods require no knowledge of the three dimensional structure of the protein. However, many of the methods of assignment do involve structural knowledge. The process of assignment is therefore also connected with methods for determining structural parameters [1]. Stage 2 of assignment can also be divided into two parts. 1) A chemical perturbation to the structure (for example modification of a residue) results in perturbation of resonances in the NMR spectrum. However, chemical perturbation does not necessarily result in spectral perturbation of the resonances of the affected group alone. This is because the local protein conformation in the region of the perturbed group may be dis-

TABLE 2

Methods for spectral assignment [1,2]

	Method	Dependent on 3D structure?
Stage 1	Physical:	
	Chemical shift Coupling pattern Coupling constant	No
Stage 2	Chemical:	
	(a) Chemical modification Sequence comparison	No
	(b) Paramagnetic probes Ionisation changes Ligand binding Solvent exchange rates	Yes

[1]Summarised in [1].
[2]Applications to lysozyme [5,7,10].

rupted. Thus, it is only possible to say in many cases
that resonances affected by a specific perturbation arise
from residues in the local region of the structure where
the perturbed group is found. 2) This is the complete
assignment and requires consideration of a whole variety
of experimental results to provide a consistent set of
assignments. This is well illustrated by the assignment
of the N(1)H protons of the six tryptophan residues of
lysozyme. The first assignments [8] were made by a variety
of techniques including chemical modification, but were
incorrect [5,9] because allowances for the more general
effects of given perturbations were not made. A relatively
straightforward example of Stage 2 assignments can be
given here. The resonances of the three tyrosine residues
of lysozyme (tyr 20, 23 and 53) are shown in Fig. 2, and
can be called T1, T2 and T3. The assignment proceeds as
follows [10]: a specific chemical modification of tyr 23
[11] perturbs resonance T1 but has no effect on T2 or T3.

As T2 and T3 are unperturbed, resonances T1 are assigned
to tyr 23. Resonances T2 are strongly perturbed by lanthanide binding whilst T1 and T3 are unaffected. Tyr 53
is close to the binding site of lanthanides, whilst tyr
20 and tyr 23 are far away. Thus resonances T2 are assigned to tyr 53 [6]. The remaining resonance (T3) must
therefore arise from tyr 20. Complete assignment of *ca* 50
resonances from *ca* 20 residues has now been possible in
lysozyme [5,7,10]. It is of particular interest that most
($\approx$ 75 per cent) of the residues close to or in the active
site region of lysozyme have one or more of their resonances assigned.

The Structure of Lysozyme

The X-ray structure of lysozyme has been determined to
a resolution of 2.5 Å [12,13]. There are several methods
by which NMR can be used to investigate features of the
solution structure. It is of interest to summarise these
methods and to compare the conclusions with those expected on the basis of the X-ray structure.

The NMR structural methods can be divided into two
groups. I shall call these groups 'qualitative' and 'quantitative'. The classification of a given method into one
or other group is based upon the existence or non-
existence of established theoretical equations which can
be used to correlate the NMR observables with structure.
The methods are summarised in Table 3. Let us consider
the qualitative methods. The first method in Table 3 is
concerned with the proximity of groups in a given region
of space. For example, lysozyme contains one histidine
residue (his 15). The ionisation of this group can be
readily followed by observing the shifts of the C(2)H and
C(4)H resonances which occur during a pH titration. It is
observed in the spectra that several other resonances
titrate with the same pK value. These shifts arise from
the local perturbation of the structure of the protein in

TABLE 3

Comparison of crystal and solution structures of lysozyme

	Method	Agreement
Qualitative	NH exchange (for tryptophans)	Fair
	Proximity of groups	Good
Quantitative	Ring current shifts	Fair
	Gd(III) Relaxation	Good
	Ln(III) Shifts	Good

the local region of his 15 [7]. In particular the resonance of an isoleucine residue, and of a valine residue
are notably perturbed. Examination of the structure shows
that ile 88 and val 92 are close to his 15. Thus there is
consistency between these NMR observations and the X-ray
structure. One observation of this type gives little information. However, the consistency of many such observations has shown that there is very good correlation between the type of groups observed to be affected by a
given perturbation, and those expected to be affected on
the basis of the X-ray structure. Consistency of this type
is important in achieving second stage assignments.

A second type of qualitative correlation is concerned
with the exchange rate of NH protons, particularly of the
tryptophan indole N(1)H protons. The rate of exchange for
surface tryptophans is expected to be greater than that
of buried tryptophans. The degree of exposure of a given
tryptophan in the X-ray structure can be estimated [8].
The rate of exchange of the different NH protons can be
followed by NMR [5,8]. The resonances of three of the
tryptophan NH protons have been assigned by chemical modification which is independent of the X-ray structure. The
observed order of exchange rate is 62 > 63 > 108 which is
the same as that predicted by the X-ray structure. Two of

the remaining tryptophans are found by NMR to exchange
slowly, and one relatively quickly. The X-ray structure
shows that two are buried (28 and 111) and one is exposed
(123). The correlation is therefore good. However, the
exchange rate of trp 123 is slower than expected relative
to trp 63 and 108. The theory of exchange is not good
enough to allow the observed results to be further interpreted.

Quantitative comparison of the structure in solution
and in the crystal is feasible in two main ways. The first
is the comparison of shifts due to aromatic ring currents,
calculated on the basis of the X-ray structure, with those
shifts observed in the assigned NMR spectrum. The latter
are taken as the observed differences between the chemical
shifts of resonances in the protein spectrum and those
found for the free amino acids. These observed differences
are called the secondary shifts. There are two major difficulties with this approach. The first is concerned with the
theory for ring current shift which is not well developed.
The second is that the assumption that the secondary shifts
arise solely from ring current effects is likely to be incorrect. It is likely that allowance must be made for
shifts caused by carbonyl or carboxylate groups, and possibly for shifts caused by other mechanisms which are not
yet understood. Nevertheless this method is potentially
a quantitative one. In Fig. 3 are plotted the observed
secondary shifts for many assigned residues against calculated [14] ring current shifts. It can be seen that there
is a correlation, but the agreement can only be described
as moderate.

The second quantitative method is concerned with spectral perturbations induced by the binding of paramagnetic
probes. The most readily interpreted effects arise from the
binding of paramagnetic lanthanide ions [3]. Lysozyme contains one major binding site for lanthanides [7, 15], in
the active site between the carboxylate groups of asp 52

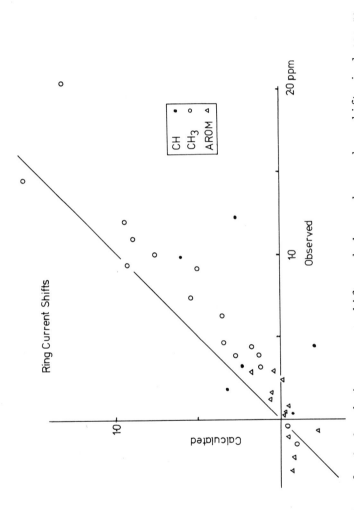

Fig. 3. Correlation of calculated ring current shifts and observed secondary shifts in lysozyme.

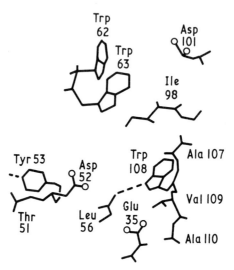

Fig. 4. Schematic illustration of the active site region of lysozyme as determined by X-ray crystallography.

and glu 35 (Fig. 4). There are a number of weaker binding sites, but the effects arising only from the strong binding site can be measured [16]. The lanthanides give rise to two types of perturbation. One of these is relaxation of resonances, induced particularly by Gd(III). The relative relaxation of resonances is simply proportional to the relative values of $1/r^6$ for the different nuclei, where r is the distance between the bound lanthanide ion and the nucleus in question. The relaxation of different resonances can be measured by the broadening of the resonances, which can be observed using difference spectroscopy even in complex regions of the spectrum [2,7,16]. Fig. 5 shows a plot of observed relative distances against the calculated relative distances from the X-ray structure [7]. Allowance has been made for rotation of methyl groups and for flipping of tyrosine residues (see below). The correlation of the NMR data with the X-ray structure is good, except at longer distances where the relaxation from the major binding site is small, and difficult to measure, and where

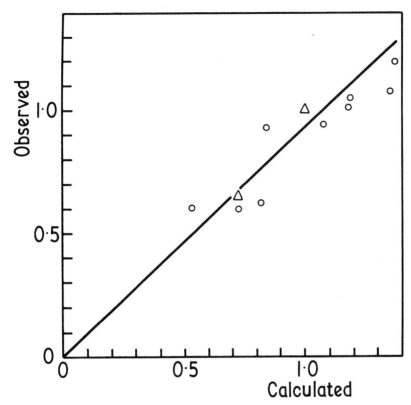

Fig. 5. Correlation of calculated and observed relative distances from the bound Gd(III) ion in lysozyme. Adapted from [7].

relaxation from weaker binding sites is more important. There are also problems associated with the exact position of the bound metal ion, and with the possibility of other motions of groups in the protein.

Paramagnetic lanthanide ions other than Gd(III) induce shifts of the resonance positions. These shifts are pro-portional to $1/r^3$, and to functions related to the angles made between the vector joining the metal to the nucleus in question and the axes of magnetic susceptibility deter-mined by the ligand field effect exerted on the bound ion by the protein. The symmetry of the ligand field cannot be determined directly. However, in many systems it appears

that an assumption of axial symmetry is a good one, parti·
cularly for Pr(III) [13]. There is evidence for approximate axial symmetry in the lysozyme-lanthanide systems
[7]. In the case of axial symmetry, the shifts are proportional to $(3 \cos^2\vartheta - 1)/r^3$, where ϑ is the angle between the metal-nucleus vector and the principal axis of
magnetic susceptibility. Using the assumption of axial
symmetry, the direction of the principal axis was varied
over all space in steps of $10°$. For a very limited range
of directions, the relative shifts of different assigned
resonances were observed [7] to be close to the relative
values of $(3 \cos^2\vartheta - 1)/r^3$. A plot of these data is given
in Fig. 6. The agreement for many resonances is excellent,
indicating that the assumption of axial symmetry is a good

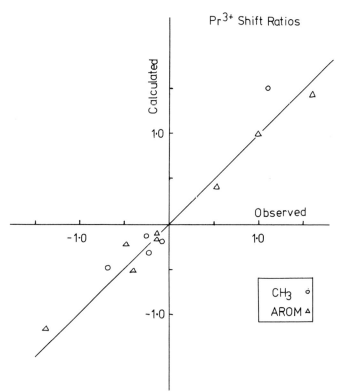

Fig. 6. Correlation of calculated and observed relative shifts induced
by the Pr(III) ion in lysozyme. An extension of a figure in [7].

one, and that the positions of the group in space is very similar in the crystal and in solution. It is exceedingly improbable that such agreement could be found if there were to be any major difference between the structures in the two phases.

To summarise, the NMR evidence in total is convincing proof of a close agreement between the solution and crystal conformations (Table 3). Further refinement of the NMR data (and the X-ray structure) is in progress, and a more detailed analysis will shortly be possible. The next section examines the evidence for mobility of groups in the lysozyme structure. It is worth mentioning that the existence of motion will make it difficult to refine the structure very far (either by NMR or by crystallography) unless the extent of the motion can be determined.

Mobility of Groups

There is increasing realization that it is incorrect to regard protein molecules as being rigid bodies. As in the previous section, comparison of evidence from NMR and crystallography will be attempted. X-ray crystallography cannot readily distinguish between 'motion' and the existence of different conformers in the crystal. Both give rise to a lack of definition in the electron density map. The NMR data show that in solution, even if there are different conformers of lysozyme present, there is rapid interconversion between them, as a single resonance is observed from each nucleus in the structure, as far as can be observed at this stage.

NMR can provide definite evidence for motion in a variety of ways, and over a wide timescale. Dipolar relaxation is sensitive to motion in the region of the resonance frequency, i.e. in the region of 10^{-9}s. T_1 and T_2 data for individual groups of lysozyme have been measured at different temperatures and at different frequencies. These data indicate the existence of rapid fluctuations of

all the residues of lysozyme [17], in other words a high
degree of internal mobility. There are, however, a number
of difficulties in interpreting the relaxation data because of the complexity of defining a theoretical model
for the motion and because of the possibility of cross-
relaxation effects [18]. It is clear that methyl groups
must rotate rapidly. Also T_2 data, measured by the Carr-
Purcell-Meiboom-Gill sequence which has recently been
applied to proteins [4,17], provides clear evidence for
extensive rapid movement of surface groups, particularly
lysine and arginine.

Evidence which can be interpreted in terms of the motion of surface groups of lysozyme in the crystal is
available, in that the electron density of many surface
lysine and arginine residues is diffuse [13]. Thus the
information from both techniques is consistent (Table 4).

TABLE 4

Mobility in Lysozyme

Method	Examples	Evidence in crystal*
Dipolar Relaxation	Lysine/Arginine (fast motion)	Yes
Equivalence of resonances	Tyrosines (flips)	No
Exchange broadening	Trp 63 and other active site resonances (oscillation)	Some
Ln(III) perturbations	Val 109 (rotation)	Some

*Data from [13]. Note 'evidence' here refers to lack of resolution.

The existence of very rapid fluctuations of all the residues of lysozyme, which the relaxation data indicate, is
consistent with the ability of aromatic rings to flip, for
this requires the ability of the protein structure surrounding these rings to be capable of movement (see below).

It is also significant that the temperature dependence of
the spectrum of lysozyme indicates that as the temperature
is increased the protein structure expands, and the ring
current shifts decrease (Campbell, Dobson and Williams,
unpublished).

Evidence for motion in NMR can be obtained by the observation of exchange effects. If an atom can exist in
chemically distinct environments, resonances corresponding
to each environment will be observed if the rate of exchange between them is slow when compared with the difference in resonance frequencies of each environment. If the
exchange rate is comparable with the difference in frequencies, line broadening will be observed. If the exchange rate is fast compared with the difference in frequencies, a single unbroadened resonance will be observed.

It has been noted above that the aromatic proton resonances of the three tyrosine residues of lysozyme are observed as two proton intensity doublets. If the tyrosine
ring were to be rigidly held in the protein, four one proton intensity resonances would be expected for each residue, as the environment of the four aromatic protons would
then, except by coincidence, be different. The observed
equivalence arises from the averaging of the chemical
shifts of the two ortho and of the two meta protons of
each residue [6]. This occurs because of rotation about
the C_β–C_γ bond of each residue. In fact, a more likely
motion is a flipping by 180° between equivalent conformations [6]. This motion has been seen for tyrosine and
phenylalanine residues in several proteins [19,20,21], and
in some cases the transition from the slow exchange to the
fast exchange situation has been studied [20]. In lysozyme
it is possible to say only that this flipping occurs. Experiments with lanthanides indicate that for at least one
residue (tyr 53) this flipping occurs at a rate of more
than 10^4 s^{-1} [6]. In the crystal this flipping cannot be
detected because the two conformations related by a 180°

rotation are indistinguishable. However it has been shown [22] that this flipping can occur only provided that other residues around the aromatic ring can 'relax' to allow the ring to rotate.

It is easy to use NMR to observe the motion of aromatic rings which flip to equivalent conformations. Evidence for the motion of other groups has been put forward. For example, the distances to the two methyl groups of val 109 in lysozyme are from the relaxation induced by Gd(III) seen to be very nearly the same (see Fig. 5). The distances predicted from the X-ray structure are different. The equivalence of the two distances would occur if val 109 were to rotate rapidly about the C_α–C_β bond [7]. The electron density of val 109 in the crystal however is not well defined [13].

It has been noted above that the C(2)H resonance of trp 63 is broad. The N(1)H resonance of this residue is also broad. Increase of temperature results in a sharpening of these lines. (Similar line broadening is observed for other resonances of the active site, but unlike that for trp 63 this is seen mainly at high pH where aggregation effects complicate the interpretation.) This line broadening is abolished in the presence of bound inhibitors. The broadening is too great to be attributed to dipolar effects, and the observed temperature effects suggest that it arises from exchange broadening which could result from the existence of relatively slow interconversion of different conformers in this region of the protein. The abolition of this effect on adding inhibitors could then be attributed to the existence of a single conformation in the bound form [23]. In the X-ray structure trp 62 (the neighbouring residue to trp 63) gives rise to diffuse electron density in the native protein, but to well defined density in the inhibitor bound form [13], showing that there is some correlation with the NMR observations.

Table 4 summarises the data discussed in this section.
There is undeniable NMR evidence for a variety of motions
in the lysozyme molecule in solution, ranging from rapid
'waving' of surface groups to slower more restricted motion of internal groups. Examination of the X-ray data in
the light of these observations shows that there is some
evidence that this motion exists even in the crystalline
state, although the timescale cannot be defined. It is of
considerable interest that this motion appears to be altered on binding of inhibitors in the active site.

A slightly different topic can be added here. The rate
of exchange of N(1)H protons of tryptophan residues has
already been mentioned. The rates of exchange of the protons of the active site groups are decreased on binding
inhibitor molecules [5,8]. This is expected for the degree
of the exposure of these groups will be reduced. However, it
is also found that the exchange rates for internal tryptophan residues *not* in the active site are reduced [5].
This implies a 'tightening' of the protein structure as
a whole on inhibitor binding, with restricted motion and
hence slower solvent exchange. Thus inhibitor binding may
affect the mobility of the structure in regions not directly associated with the binding.

Conformational Changes in Lysozyme

One of the most interesting aspects of protein structure is concerned with the response of the structure to
chemical perturbations such as binding of ions or ligands,
or chemical modifications. In this article I shall concentrate on a particular region of the structure, that of
the region of the catalytic carboxylate group of glu 35
in the X-ray structure. The proposed lysozyme mechanism
[13] involves proton transfer from glu 35, therefore the
effects of ionisation of this group are of considerable
interest.

The effect of pH changes on the spectrum of lysozyme

produces surprising shifts of the resonances of trp 108.
This is clearly shown up by using difference spectroscopy
(Fig. 7). Fig. 7 shows that in the aromatic region of the
lysozyme spectrum, two resonances shift markedly upfield

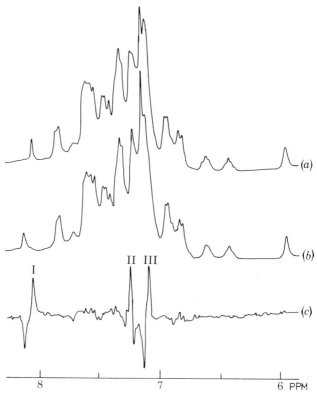

Fig. 7. Spectra of 5 mM lysozyme at 54°C. (a) pH 6.5, (b) pH 6.0,
(c) difference (a)-(b) on twice the vertical scale. The resonances
I and III are from his 15, and II is from the C(2)H proton of trp
108. Taken from [9].

as the pH is increased. These arise from the protons of
his 15 which titrate in the expected manner. Another resonance, that assigned by chemical modification to the
C(2)H proton of trp 108, shifts downfield with increasing
pH. The N(1)H proton of trp 108 shifts in a similar manner.
The data are plotted in Fig. 8. The pK value obtained from
this titration is 6.2 ± 0.1 [9], and this is the value for

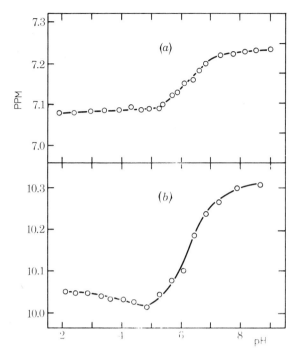

Fig. 8. pH dependence of the chemical shift values of the resonances of trp 108; (a) C(2)H proton, (b) N(1)H proton. 5 mM lysozyme, 54°C. Taken from [9].

glu 35. This has been confirmed by examining the effects of metal binding to glu 35 [7]. Before drawing conclusions from this, it is of interest to discover whether this observation is a general feature of lysozymes. The lysozyme from human sources differs most in sequence from hen lysozyme (differences in *ca* 40 per cent of the residues) [13]. However Figs. 9 and 10 show that analogous behaviour of the conserved trp 108 in human lysozyme is observed. The pK of glu 35 in this protein can be measured as 7.2 ± 0.1 under these conditions [9]. It is interesting that the pK is higher in human lysozyme than in hen, and that the shifts on trp 108 are larger (Figs. 8, 10). Other changes in the spectrum indicate that a small conformational change takes place for other groups in the region of trp 108.

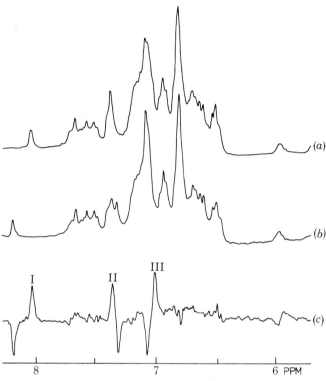

Fig. 9. Spectra of 5 mM human lysozyme, 54°C. (a) pH 7.0, (b) pH 6.5, (c) difference (a)-(b) on twice the vertical scale. The resonances I and III are from his 77, and II is from the C(2)H proton of trp 108. Taken from [9].

It is now possible to ask the reason for the observed NMR shifts in these proteins. Shifts of this type have not been observed previously, and as the groups involved are in the active site, and the interaction appears in different lysozymes, it is reasonable to suppose that a specific interaction exists between the carboxylate group and the tryptophan ring. This interaction is pH dependent. If we suppose that in its neutral state only, the carboxylic acid interacts with the tryptophan ring, then on ionisation moves away from the tryptophan ring, the NMR observations can be explained. The movement of the carboxylate group will cause a change in the anisotropic shielding of

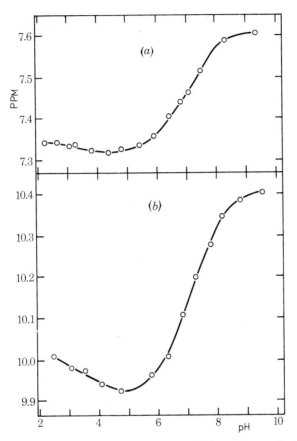

Fig. 10. pH dependence of the chemical shift values of the resonances of trp 108 in human lysozyme; (a) C(2)H proton, (b) N(1)H proton. 5 mM human lysozyme, 54°C. Taken from [9].

the tryptophan protons by the carboxylate group and result in a shift. The interaction of the protonated form only explains the high pK value of the group [9]. This is an example of a group in the active site being in a special environment to enable catalytic action to occur. The exact nature of the interaction cannot be defined by the NMR experiment. An hydrogen bonding system is one possibility, but a charge transfer mechanism is also conceivable.

In the presence of the bound inhibitor N-acetyl glucosamine (GlcNAc), the pH behaviour of trp 108 is altered,

C.M. DOBSON

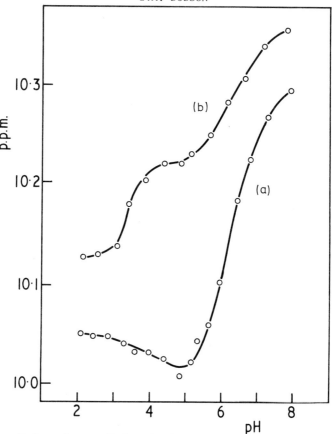

Fig. 11. pH dependence of the chemical shift values of the N(1)H proton resonance of trp 108; (a) native protein, (b) protein with 0.22 M GlcNAc. 5 mM lysozyme, 54°C.

Fig. 11. It is interesting that the chemical shift of the trp 108 resonances are now perturbed both by glu 35 ionisation and by asp 52 ionisation. The latter group appears to have a pK value of 3.3 ± 0.2. Presumably the ionisation of asp 52 affects the relative orientation of glu 35 and trp 108.

An observation of the type discussed above can be obtained rapidly by NMR. No data are available from X-ray crystallography, because the pH cannot readily be varied in the crystal (because of difficulties with change in the unit cell characteristics) and because small changes of specific groups are difficult to follow. Equally, compari-

son of the behaviour of different, but related, proteins
is rapid by NMR but difficult by crystallography.

It is however possible to cause a greater interaction
between glu 35 and trp 108, and to compare the NMR data
with X-ray data. Chemical modification of trp 108 with
iodine results in the formation of an ester between glu 35
and trp 108 [24]. The existence of this ester was elegantly
shown by X-ray diffraction [25,26], and can be observed
from the NMR spectrum [5]. The spectrum of the methyl
group resonances and of the aromatic protons shows that
trp 108 must have moved, as a result of this modification,
for the resonances of groups such as met 105, ile 98 are
perturbed [5]. These groups are ring current shifted by
trp 108. However, a more interesting observation is that
the resonances of leu 17 and of tyr 23 (Fig. 12) are also
shifted (note that the resonances of tyr 23 were assigned
by chemical modification). Examination of the crystal

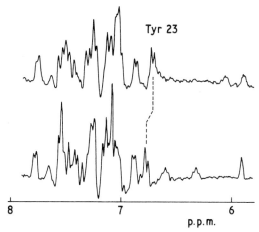

Fig. 12. Convolution difference spectra of the aromatic protons of
(a) native lysozyme, (b) lysozyme modified at trp 108. The shift of
the *ortho* proton resonance of tyr 23 is indicated.

structure shows these groups are distant from 108 (*ca* 15Å),
and that shifts could not arise solely from trp 108 movement. Instead, the conformational change must have been
transmitted from trp 108 through the 'hydrophobic box' of

trp 111 and trp 28 to tyr 23. This is an important observation for two reasons. First, it shows that conformational changes can be transmitted through a protein. Such
a transmission is expected to occur in allosteric systems.
Secondly, the X-ray data suggest that the conformational
changes in lysozyme when the modification is carried out
in the crystal are localised to the movement of glu 35 and
a slight movement of trp 108 [25]. This is not in agreement with the NMR observations.

Rate Processes

If a conformational change results in shifts of resonance positions, it is possible to use NMR to provide information about the rate of the conformational change. Suppose the conditions are set so that the solution contains
lysozyme in two conformations, the native and the perturbed
conformation. If a nucleus has a different chemical shift
in the two conformations then the appearance of the spectrum will depend upon the rate of interconversion of the
two forms, in just the same manner as the tyrosine resonances depend upon the rate of flipping.

The conformational change dependent upon ionisation of
the glu 35 carboxylate group causes a continuous shift of
the trp 108 resonances as the pH is varied. No line broadening is observed on the resonances, showing that the conformational change is fast compared to the difference of
chemical shifts in the high and low pH forms. The rate of
this conformational change is therefore greater than *ca*
10^3 s^{-1}.

A conformational change about which more detailed information can be extracted is that induced by binding of inhibitors. The binding of inhibitors $(GlcNAc)_n$ where n =
1-3 has been studied [8,23]. The spectral changes are similar for each inhibitor suggesting that binding in site C
(the only site occupied by all these inhibitors) is of
major importance for inducing the conformational change.

In this section I shall concentrate again on a region of
the protein close to trp 108. A peak at ~ 6.2 ppm is
strongly shifted by inhibitor binding. This peak arises
from the protons of the benzoid ring of a tryptophan [1].
This peak is very markedly shifted on trp 108 modification
(see Fig. 12) which implies that the tryptophan residue
from which it arises is in the region of trp 108. Ring
current shift calculation [14] suggest that it is trp 28,
but the assignment is not certain. For our purpose it is
only necessary to observe that it is shifted by inhibitor
binding, and likely to be close to trp 108.

Binding of (GlcNAc)$_3$ to the protein at low temperatures
shows that separate resonances of this proton are observed
in the native and inhibitor bound protein, i.e. the exchange is slow [23]. A titration is shown in Fig. 13. When
a concentration of inhibitor is chosen such that a mixture
of 50 per cent of bound and unbound forms exists, the spectra of Fig. 14 are obtained. These were recorded at different temperatures, and the change from slow through
intermediate to fast exchange is clearly observed. From
these spectra it is possible to measure the role of the
conformational change giving rise to the induced shifts,
by measuring the line broadening which occurs.

The rate constant obtained is $1.3 \times 10^2 \text{ s}^{-1}$ at 45°C
[23]. The kinetics of the binding of (GlcNAc)$_3$ have been
studied by chemical methods, and involve a rapid prequili-
brium step followed by a slower step [27,28,29,30], i.e.

$$E + I \underset{k_{-1}}{\overset{k_1}{\rightleftarrows}} EI_1 \underset{k_{32}}{\overset{k_{23}}{\rightleftarrows}} EI_2$$

The rate constant k_{32} corresponds most nearly to that
measured by NMR. the other rate constants are faster.
Thus, the NMR data show that the slow step in the binding
is a conformational change in the active site. In agreement with chemical studies [13], the NMR data showed that
with GlcNAc the conformational change was more rapid.

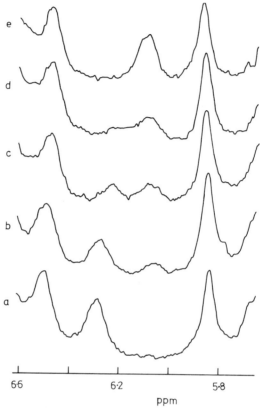

Fig. 13. Part of the spectrum of 8 mM lysozyme at 37°C, pH 5.3 in the presence of different concentrations of (GlcNAc)$_3$ which are (a) 0 mM; (b) 1.9 mM; (c) 4.2 mM; (d) 5.1 mM; (e) 7.7 mM. The proton giving rise to the resonance at *ca* 6.2 ppm is observed to exist in two environments. From [23].

Conclusions

This article has been concerned with a variety of features of the structure of an enzyme in solution. It is possible to summarise the correlation between the NMR studies of lysozyme in solution, and the X-ray diffraction studies of crystalline lysozyme.

The first conclusion is that the 'time-averaged' positions of groups in the solution structure are very close to those determined by X-ray diffraction. There is a good

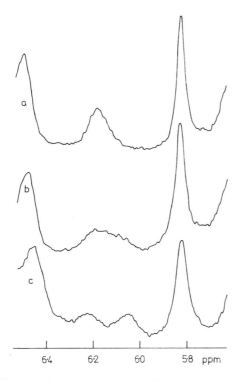

Fig. 14. The effect of temperature on the spectrum of 8 mM hen lysozyme containing 4.1 mM (GlcNAc)$_3$ at pH 5.3. Temperatures are (a) 55°
C; (b) 45°C; (c) 37°C. The passage from slow through intermediate to
fast exchange conditions is observed. Corrected from [23].

correlation between observed NMR data and expectations
based on the X-ray structure.

The second conclusion is that groups in the protein
structure have a degree of independent mobility in solution. This mobility is extensive on the surface, and somewhat more restricted in the interior of the protein. The
lack of definition in the electron density map for certain
groups in the crystal is consistent with many of the NMR
results. It is of particular interest that the mobility
of groups in the active site appears, from both NMR and
X-ray data, to be more restricted when inhibitors are
bound.

The third conclusion concerns conformational changes

which occur when the molecule is perturbed. NMR can readily
detect the existence of conformational changes, and define
the groups involved. X-ray crystallography can define the
extent of the changes more quantitatively. However, it
seems possible that there are differences between the conformational changes which occur in the crystal and in solution. For example, the modification of trp 108 appears to
have more extensive effects on the solution structure than
on the crystal structure. Related to this point is the observation that the conformational changes in different
crystal forms are not identical. The reason for this may
be that in a crystal, conformational changes are resisted
by the molecular packing. Thus, when a partial perturbation is applied in the crystal, the effects may be different from those experienced when the perturbation is applied
in solution.

Overall, therefore, it may be expected that in general
X-ray crystallography will give at least a good first approximation to the description of solution structures of
globular proteins. However, more studies are required to
interpret details and the effects of induced conformational changes carried out in crystals. This aspect of conformational behaviour is of particular importance when the
mechanism of action of the protein involves such conformational changes, which will be more important in allosteric
systems.

Acknowledgements

This article summarises work carried out in conjunction
with I.D. Campbell, R. Cassels, F.M. Poulsen and R.J.P.
Williams. We thank C.C.F. Blake, L.N. Johnson and D.C.
Phillips for useful discussions of the crystallographic
data referred to here, and S.J. Perkins for providing the
ring current calculations prior to publication.

References

1. Campbell, I.D., Dobson, C.M. and Williams, R.J.P. (1975). *Proc. R. Soc.* **A 345**, 23.
2. Campbell, I.D., Dobson, C.M., Williams, R.J.P. and Xavier, A.V. (1973). *J. Magnetic Resonance* **11**, 172.
3. Dobson, C.M. and Levine, B.A. (1976). *In: New Techniques in Biophysics and Cell Biology* (ed. Pain, R.H. and Smith, B.E.), Vol. III, p. 19, Wiley.
4. Campbell, I.D., Dobson, C.M., Williams, R.J.P. and Wright, P.E. (1975). *FEBS Lett.* **57**, 96.
5. Cassels, R., Dobson, C.M., Poulsen, F.M. and Williams, R.J.P. (1977). In preparation.
6. Campbell, I.D., Dobson, C.M. and Williams, R.J.P. (1975). *Proc. R. Soc.* **B189**, 503.
7. Campbell, I.D., Dobson, C.M. and Williams, R.J.P. (1975). *Proc. R. Soc.* **A345**, 41.
8. Glickson, J.D., Phillips, W.D. and Rupley, J.A. (1971). *J. Amer. Chem. Soc.* **93**, 4031.
9. Campbell, I.D., Dobson, C.M. and Williams, R.J.P. (1975). *Proc. R. Soc.* **B189**, 485.
10. Dobson, C.M., Ferguson, S.J., Poulsen, F.M. and Williams, R.J.P. (1977). In preparation.
11. Aboderin, A.A. and Boedefeld, E. (1976). *Biochim. Biophys. Acta* **420**, 177.
12. Blake, C.C.F., Johnson, L.N., Mair, G.A., North, A.C.T., Phillips, D.C. and Sarma, V.R. (1967). *Proc. R. Soc.* **B167**, 378.
13. Imoto, T., Johnson, L.N., North, A.C.T., Phillips, D.C., and Rupley, J.A. (1972). *In: The Enzymes*, Vol. III (3rd edn.) (ed. Boyer, P.D.), Academic Press, N.Y., 666.
14. Perkins, S.J. (1977). Personal communication.
15. Morallee, K.G., Nieboer, E., Rossotti, F.J.C., Williams, R.J.P., Xavier, A.V. and Dwek, R.A. (1970). *J.C.S. Chem. Comm.* 1132.
16. Dobson, C.M. and Williams, R.J.P. (1977). *In: Metal-Ligand Interactions in Organic Chemistry and Biochemistry* (ed. Pullman, B. and Goldblum, N.), Part I, p. 255, Reidel.
17. Campbell, I.D., Dobson, C.M., Ratcliffe, R.G. and Williams, R.J.P. (1977). In preparation.
18. Kalk, A. and Berensen, H.J.C. (1976). *J. Mag. Res.* **24**, 343.
19. Wüthrich, K. and Wagner, G. (1975). *FEBS Lett.* **50**, 265.
20. Campbell, I.D., Dobson, C.M., Moore, G.R., Perkins, S.J. and Williams, R.J.P. (1976). *FEBS Lett.* **70**, 96.
21. Cave, A., Dobson, C.M., Parello, J. and Williams, R.J.P. (1976). *FEBS Lett.* **65**, 190.
22. Gelin, B.R. and Karplus, M. (1975). *Proc. Nat. Acad. Sci. U.S.A.* **72**, 2002.
23. Dobson, C.M. and Williams, R.J.P. (1975). *FEBS Lett.* **56**, 362.
24. Imoto, T. and Rupley, J.A. (1973). *J. Molec. Biol.* **80**, 657.
25. Beddell, C.R. and Blake, C.C.F. (1970). *In: Chemical Reactivity and Biological Role of Functional Groups of Enzymes* (Smellie, R.M.S., ed), p. 157, Academic Press, N.Y.
26. Beddell, C.R., Blake, C.C.F. and Oatley, S.J. (1975). *J. Molec. Biol.* **97**, 643.
27. Holler, E., Rupley, J.A. and Hess, G.P. (1969). *Biochem. Biophys.*

   Res. Comm. **37**, 423.
28. Holler, E., Rupley, J.A. and Hess, G.P. (1970). *Biochem. Biophys.*
   Res. Comm. **40**, 166.
29. Sykes, B.D. (1969). *Biochemistry* **8**, 1110.
30. Sykes, B.D. and Parravano, C. (1969). *J. Biol. Chem.* **244**, 3900.
31. Roberts, G.C.K., Feeney, J., Birdsall, B., Kimber, B., Griffiths,
   D.V., King, R.W. and Burgen, A.S.V. (this volume).
32. McDonald, C.C. and Phillips, W.D. (1969). *Biochem. Biophys. Res.*
   Comm. **35**, 43.

DIHYDROFOLATE REDUCTASE:
THE USE OF FLUORINE-LABELLED AND SELECTIVELY DEUTERATED ENZYME TO STUDY SUBSTRATE AND INHIBITOR BINDING

G.C.K. ROBERTS, J. FEENEY, B. BIRDSALL, B. KIMBER,
D.V. GRIFFITHS,* R.W. KING and A.S.V. BURGEN

Division of Molecular Pharmacology,
National Institute for Medical Research, Mill Hill, London NW7 1AA

Introduction

Dihydrofolate reductase catalyses the reduction of dihydrofolate to tetrahydrofolate, using NADPH as coenzyme. Tetrahydrofolate plays a vital role in intermediary metabolism, acting as a 'carrier' of one-carbon fragments in the biosynthesis of several amino-acids, thymidylate and purines (for a review see Blakley [1]). In most of these one-carbon transfer reactions, the folate coenzyme remains at the tetrahydro oxidation level, but in the synthesis of thymidylate from deoxyuridylate (catalysed by thymidylate synthetase) the product is dihydrofolate. This must then be reduced by dihydrofolate reductase in order to maintain the cellular pool of tetrahydrofolate compounds. Inhibition of dihydrofolate reductase will thus lead to a depletion of this pool, and to inhibition of DNA synthesis due to lack of thymidylate and purines.

Over the last thirty years a large number of inhibitors of dihydrofolate reductase have been synthesised, and several of these have found clinical application [1,2,3, 4]. The structures of the substrates, folate and dihydrofolate, and of two representative inhibitors, methotrexate and trimethoprim, are shown in Fig. 1. Methotrexate is

---

Present address: Dept. of Chemistry, University of Keele.

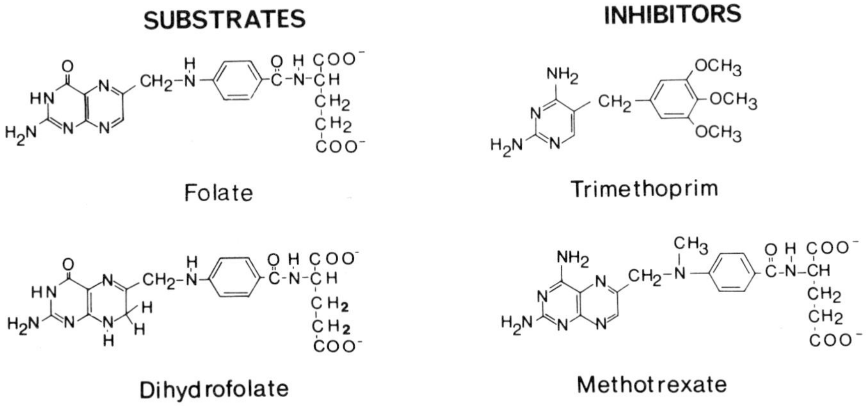

SUBSTRATES   **INHIBITORS**

Folate   Trimethoprim

Dihydrofolate   Methotrexate

Fig. 1. The structures of folate, dihydrofolate, trimethoprim and methotrexate.

clearly a close structural analogue of folate, the essential difference being the replacement of the 4-oxo substituent on the pteridine ring by an amino group (methylation of N_{10} has little effect on the binding to the enzyme). This change in the 4-substituent leads to an increase of approximately 10,000-fold in the binding constant for dihydrofolate reductase, and all the most effective inhibitors have an amino group in the 4-position. Even trimethoprim, which retains only the 2,4-diaminopyrimidine ring, binds a hundredfold more tightly than does folate.

We have been studying the dihydrofolate reductase from *Lactobacillus casei* MTX/R [5] by a variety of techniques, including NMR spectroscopy, in an attempt to understand in detail how substrates, inhibitors and coenzymes bind to this enzyme. In particular we are seeking a molecular explanation of the dramatic effect of the 4-substituent on the binding constant.

In common with the enzyme from other sources, the dihydrofolate reductase from *L. casei* is a small (Mol. wt. 18,000) monomeric protein. Even so, in the aromatic region of the ^{1}H NMR spectrum the only resonances from individual

amino-acid residues which can clearly be resolved are
those of the histidine residues (Fig. 2). Although there
are obviously extensive changes in the main aromatic region of the spectrum when trimethoprim or methotrexate
bind to the enzyme (Fig. 2), we cannot describe these

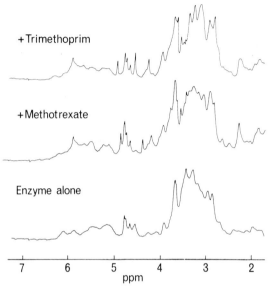

Fig. 2. The aromatic region of the 270 MHz ^{1}H NMR spectrum of L.
$casei$ dihydrofolate reductase alone (bottom) and in the presence of
one molar equivalent of methotrexate or trimethoprim.

changes in terms of the effects of inhibitor binding on
individual residues — clearly a prerequisite for any detailed understanding of ligand binding. It is clear that
we must find ways of simplifying the spectrum if we are to
be able to obtain this kind of information. We have used
two approaches to achieve this simplification: the preparation of enzyme containing fluorine-substituted aromatic
amino-acids, and the preparation of selectively deuterated
enzymes. These have allowed us to study the effects of
ligand binding on individual amino-acid residues in considerable detail; with their help we have been able to
examine some 20 residues in the protein. A brief survey of

our results with these 'substituted' enzymes is given
here; more detailed descriptions are given in refs. [6–8].

Fluorine-Labelled Enzymes

We have succeeded in preparing dihydrofolate reductase
containing 6-fluorotryptophan, 3-fluorotyrosine or m-
fluorophenylalanine in place of the corresponding normal
amino-acids. Although the enzyme yields are low, and the
fluorine-labelled enzymes are significantly less stable
than the native enzyme, their specific activity is essentially normal [8].

The ^{19}F spectrum of the 6-fluorotryptophan-substituted
enzyme is shown in Fig. 3; there are five tryptophan residues in the protein, and their ^{19}F resonances appear over
a range of 5.6 ppm. Three resonances appear close to the

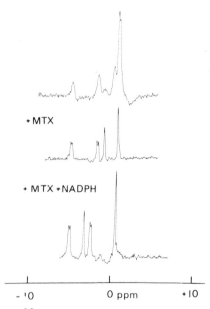

Fig. 3. The 94.1 MHz ^{19}F NMR spectrum of 6-fluorotryptophan-
containing dihydrofolate reductase alone (top) and in the presence of
one molar equivalent of methotrexate or methotrexate plus NADPH. The
spectra were obtained under conditions of gated proton noise decoupling (gated on only during data acquisition, and off during a delay
before the next pulse). The zero on the chemical shift scale is set
at the resonance position of the denatured enzyme.

position of the 6-fluorotryptophan resonance in the denatured enzyme, and presumably represent tryptophan residues accessible to the solvent. The other two resonances
appear 1.3 and 4.5 ppm to low field. In considering possible mechanisms for these large differences in chemical
shift, the most likely candidate [8,9] seems to be the
second-order electric field shift, or van der Waals shift,
where the shift arises from the same fluctuating electric
fields that give rise to London dispersion forces. On this
hypothesis one would expect that the ^{19}F resonances of
'buried' fluoroamino-acids in proteins would appear to low
field of the resonance position for the denatured protein,
and this in general seems to be what is observed.

On addition of methotrexate, there is a general sharpening of all the ^{19}F resonances, revealing that the two at
lowest field are 'doublets'. This splitting does not arise
from ^{1}H–^{19}F spin–spin coupling, since the spectra were obtained under conditions of proton noise-decoupling. The
two other obvious possibilities are a chemical shift difference, implying that these two tryptophan residues each
exist, with approximately equal probability, in two conformational states, or a ^{19}F–^{19}F through-space coupling.
The necessary experiments to distinguish between these ex-
planations are in progress. As well as the line-sharpening,
addition of methotrexate leads to changes in chemical
shift of three of the five fluorotryptophan resonances.
The two 'doublets' show small downfield shifts, and one of
the three resonances from 'solvent-accessible' tryptophans
shows a large downfield shift (1.1 ppm; see Fig. 3). Subsequent addition of NADPH to form the ternary complex
affects the same three resonances; in particular, the resonance which undergoes a large downfield shift on methotrexate binding is further affected by coenzyme, appearing
at 3.5 ppm (Fig. 3). Thus both inhibitor and coenzyme
affect a single tryptophan residue; if the shifts do arise
from the van der Waals mechanism, the downfield direction

would be consistent with a direct interaction between the
ligands and this tryptophan residue. There is chemical
evidence (K. Hood and G.C.K. Roberts, unpublished work)
that a tryptophan is involved in the coenzyme binding
site.

Fig. 4 shows the ^{19}F spectrum of the 3-fluorotyrosine-
substituted dihydrofolate reductase; the resonances of the
five fluorotyrosines appear over a range of 2.7 ppm, one
(K) being 2 ppm to low field of the position of the denatured enzyme. It is interesting to note that the corresponding ^{1}H resonances (the 3,5-protons, observed in a

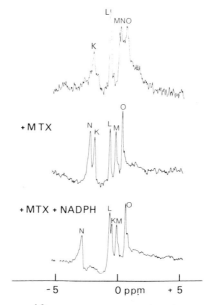

Fig. 4. The 94.1 MHz ^{19}F NMR spectrum of 3-fluorotyrosine-containing
dihydrofolate reductase alone (top) and in the presence of methotrexate or methotrexate plus NADPH. Conditions as for Fig. 3.

selectively deuterated enzyme, see below) appear over a
range of only 0.3 ppm, only a tenth of the shifts in the
^{19}F spectra. Once again, the relatively broad lines (15–
30 Hz) in the spectrum of the free enzyme sharpen markedly
on addition of ligands (Fig. 4). This sharpening is also
seen in many regions of the ^{1}H spectrum, and the most

probable explanation seems to be that the enzyme exists in
a number of interconverting conformations, while the lig-
and complexes are essentially in a single conformation.
There is some supporting evidence for this from stopped-
flow studies (R.W. King, S. Dunn, J.G. Batchelor, G.C.K.
Roberts and A.S.V. Burgen, unpublished work). The major
effect of the addition of methotrexate (Fig. 4) is a 2 ppm
downfield shift of one fluorotyrosine resonance, N; the
same resonance is shifted another 1.3 ppm downfield on
subsequent addition of coenzyme. Interestingly, coenzyme
binding leads to an *upfield* shift of resonance K by 1.5
ppm; the largest shift in the ^{1}H spectrum is 0.2 ppm. This
comparison rules out neighbour magnetic anisotropy as the
origin of the shift of the ^{19}F resonance.

The simple proximity of the bound ligand to the fluorotryptophan residue giving rise to this resonance would be
expected to lead to a *downfield* shift, from the van der
Waals mechanism, as discussed above. If the observed
shift does arise from the van der Waals mechanism, an
upfield shift can only result from a group or groups on
the enzyme moving further away from the fluorotryptophan
– that is from a coenzyme-induced conformational change in
the protein. The details of coenzyme binding are discussed
in the following paper in this volume.

The large changes in chemical shift seen in the ^{19}F
spectra on ligand binding (discussed in detail in ref.[9])
are clearly an advantage. However, fluorine substitution
obviously represents a non-trivial change in the chemistry
of the protein. For example, we have some evidence that
the fluorotryptophan-substituted enzyme has a lower affinity for coenzyme than does the native enzyme. Some cau-
tion must undoubtedly be exercised in applying the results
from fluorine-substituted enzymes to the normal enzyme.

Selectively Deuterated Enzymes

In contrast to fluorine-substitution, deuteration would
be expected to have minimal effects on the properties of
the protein. We have prepared a series of selectively
deuterated analogues of dihydrofolate reductase, in which
the bulk of the aromatic protons have been replaced by
deuterium so as to simplify the aromatic region of the ^{1}H
spectrum of the protein. The spectrum of the first of
these, in which the only proton resonances remaining in
the aromatic region are those of the 2,6-protons of the
tyrosine residues [7], are shown in Fig. 5. The simplification arising from reducing the number of aromatic protons from 91 to 10 is apparent; in the selectively deuterated enzyme, the effects of ligands on the tyrosine residues can be studied in detail.

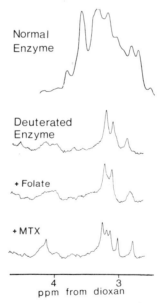

Fig. 5. The aromatic region of the 270 MHz ^{1}H NMR spectrum of (top)
isotopically normal *L. casei* dihydrofolate reductase, and the selectively deuterated enzyme in which the only remaining aromatic proton
resonances are those of the 2,6 protons of the tyrosine residues.
Spectra of the complexes of the selectively deuterated enzyme with
methotrexate and with folate are also shown.

As can be seen in Fig. 5, addition of folate only affects the chemical shift of the highest-field tyrosine
resonance. Methotrexate, however, affects two other tyrosine resonances as well. Of the 16 individual aromatic
residues we have so far studied, nine are affected by the
binding of substrates or inhibitors; of the nine, only
three are affected in a similar way by both folate and
methotrexate [6,7,8]. This suggests that there are fairly
extensive differences in the mode of binding of substrates
and inhibitors.

The majority of the substrates and inhibitors bind relatively tightly to dihydrofolate reductase ($K_a > 10^{-5}$ M^{-1}),
and slow exchange conditions prevail for both the ^{19}F and
the ^1H spectra of the complexes. This leads to a problem
of assignment — how can we connect the resonances in the
spectra of the substrate or inhibitor complexes with those
in the spectra of the free enzyme? The approach we have
most often used (for both ^1H and ^{19}F spectra) is to study
the effects of 'fragments' of the inhibitors, which bind
considerably more weakly and are thus usually in fast
exchange. The 'fragments' of methotrexate we have studied,
whose structures are shown in Fig. 6, are 2,4-diamino-
pyrimidine and p-amino- or p-nitro-benzoyl-L-glutamate.
As one would anticipate from their structural relationship
to methotrexate, these 'fragments' are able to bind to the
enzyme simultaneously. Fig. 7 shows the effects of adding

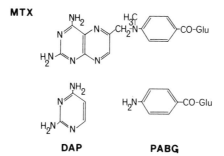

Fig. 6. Structures of methotrexate (MTX) and its 'fragments', 2,4-
diaminopyrimidine (DAP) and p-amino-benzoyl-L-glutamate (PABG).

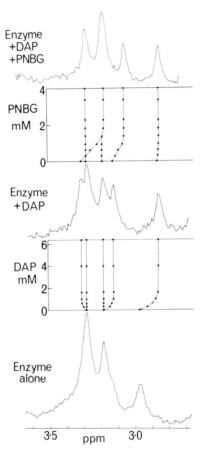

Fig. 7. The effects of adding increasing concentrations of 2,4-diaminopyrimidine (DAP), followed by p-nitro-benzoyl-L-glutamate (PNBG), on the resonances of the tyrosine 2,6-protons of selectively deuterated dihydrofolate reductase.

first 2,4-diamino-pyrimidine and then p-nitrobenzoyl-L-glutamate on the resonances of the 2,6-protons of the tyrosine residues. It is apparent that both ligands are in fast exchange, so that progressive shifts of the tyrosine resonances are observed as the ligand concentration is increased. The resonances of these complexes can therefore be connected with those of the free enzyme in a straightforward manner.

The spectrum of the enzyme-2,4-diaminopyrimidine-p-

nitrobenzoyl-L-glutamate complex is closely similar to
that of the enzyme-methotrexate complex. The position of
the tyrosine resonances in these complexes are compared in
Fig. 8; one tyrosine resonance differs in chemical shift

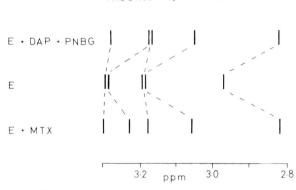

TYROSINE 2,6-H

Fig. 8. A comparison of the chemical shifts of the resonances of the
2,6-protons of the tyrosine residues of dihydrofolate reductase in
the free enzyme (centre), the enzyme-2,4-diaminopyrimidine-p-nitro-
benzoyl-L-glutamate complex (top) and the enzyme-methotrexate complex
(bottom).

by approximately 0.05 ppm between the two complexes, but
otherwise the agreement is very good, and a similar degree
of correspondence is seen in other regions of the spectrum.
We can thus conclude that the 'fragments' bind to the enzyme in a way which is closely similar to the binding of
the corresponding moieties of methotrexate, and we are
therefore justified in using the results with the 'fragments' (e.g. Fig. 7) to connect the resonances in the
spectrum of the enzyme-methotrexate complex with those in
the free enzyme, as indicated in Fig. 8.

    The use of 'fragments' of the ligands of interest has
another potential advantage. It should allow one to establish which of the tyrosine residues is affected by which
part of the ligand molecule. However, in the present case
this proves to be difficult to do, though for a very interesting reason. Not only do 2,4-diaminopyridine and p-
aminobenzoyl-L-glutamate bind *simultaneously* to the

enzyme, they also bind *cooperatively*, in the sense that
p-aminobenzoyl-L-glutamate binds fifty times better to the
enzyme-2,4-diaminopyrimidine complex than it does to the
enzyme alone. The binding constants, determined by fluori-
metric methods (B. Birdsall, J. de Miranda, A.S.V. Burgen
and G.C.K. Roberts, unpublished work), are given in Table
1.

TABLE 1

*Binding constants for 2,4-diaminopyrimidine and p-aminobenzoyl-
L-glutamate to* L. casei *dihydrofolate reductase*

Reaction	$10^{-3} \times K_a (M^{-1})$	$\Delta G(kJ/mole)$
E + DAP $\longrightarrow$ E.DAP	1.27	17.6
E + PABG $\longrightarrow$ E.PABG	0.83	16.6
E.DAP + PABG $\longrightarrow$ E.DAP.PABG	49.6	26.7
E.PABG + DAP $\longrightarrow$ E.DAP.PABG	60.2	27.1

The most likely explanation (for which there is supporting evidence from other experiments) for this cooperativity is that the binding of 2,4-diaminopyrimidine or p-
aminobenzoyl-L-glutamate induces a conformational change
which increases the affinity of the enzyme for the other
ligand. If this is indeed happening, one would expect that
some of the changes in chemical shift of resonances arising from the protein which occur when these ligands bind
would arise from this conformational change, rather than
from a direct effect of the ligand on residues in its
immediate vicinity. The chemical shifts (or pK values) of
those histidine and tyrosine residues which are affected
by the binding of 2,4-diaminopyridine or p-amino- or p-
nitrobenzoyl-L-glutamate are collected in Table 2.
   Consider first histidine H_F, which shows the simplest
behaviour; its pK is unaffected by the binding of 2,4-

TABLE 2

Effects of 2,4-diaminopyrimidine, and p-amino or p-nitro-
benzoyl-L-glutamate on the tyrosine (Y) and Histidine (H)
residues of L. casei *dihydrofolate reductase*

Complex	$H_A(\delta)$	$H_E(pK)$	$H_F(pK)$	$Y_A(\delta)$	$Y_D(\delta)$	$Y_E(\delta)$
E	4.79	6.54	6.54	3.29	3.19	2.97
E.DAP	4.90	6.14	6.54	3.33	3.13	2.82
E.PABG ⎫	4.89	6.28	~7.2			
E.PNBG ⎭				3.23	3.19	2.85
E.DAP.PABG ⎫	4.93	6.15	~7.2			
E.DAP.PNBG ⎭				3.17	3.05	2.82

diaminopyridine, but increases approximately 0.7 units on
the binding of p-aminobenzoyl-L-glutamate, and this latter
effect is the same, whether 2,4-diamopyrimidine is present
or not. However, the chemical shifts of the resonances of
histidine H_A and tyrosine Y_E are affected more or less
equally by the binding of either *or both* fragments. This
behaviour is most readily explained if these chemical
shift changes are conformational in origin — indeed both
these resonances are also affected by coenzyme binding.
In the case of tyrosine Y_D, while p-nitrobenzoyl-L-
glutamate itself has no effect, it almost doubles the
effect of 2,4-diaminopyrimidine; once again a conformational mechanism seems most probable. We thus conclude
that some, at least, of the changes in chemical shift we
observe when these ligands bind have their origin in a
conformational change, which could well be related to the
conformational change responsible for the cooperativity in
ligand binding. We have also studied the binding of a number of structural analogues of 2,4-diaminopyridine and p-
amino-benzoyl-L-glutamate, and find that the degree of cooperativity is a sensitive function of the structure of
the ligands. We are now attempting to relate the degree of

cooperativity displayed by these various ligands to the
nature of the conformational change as manifested in the
NMR spectra (B. Birdsall, J. de Miranda, A.S.V. Burgen,
G.C.K. Roberts and J. Feeney, unpublished work).

It was shown above that the spectrum of the enzyme-
methotrexate complex is very similar to that of the enzyme-
2,4-diaminopyrimidine — p-nitrobenzoyl-L-glutamate complex;
since it appears that some of the chemical shift changes
in the latter case arise from conformational changes, this
must also be true in the former. The cooperativity must
thus be present in methotrexate binding too, though here
it is 'cooperativity' between two parts of the same molecule. This phenomenon clearly has important implications
for attempts to understand, in molecular terms, the relationship between structure and activity among inhibitors
of dihydrofolate reductase. It seems likely that the study
of 'fragment' binding and cooperativity, allied to kinetic
and NMR experiments, will afford a valuable opportunity to
relate the structural, dynamic and energetic aspects of
these conformational changes to one another.

Acknowledgements

We are most grateful to Ms. P. Scudder and Ms. G.
Ostler for their skilful technical assistance, to the
Trustees of the Beit Memorial Fund for a fellowship to
D.V.G., and to the Wellcome Foundation for financial sup-
port for B.B.

References

1. Blakley, R.L. (1969). *The biochemistry of folic acid and related
   pteridines.* North-Holland, Amsterdam.
2. Baker, B.R. (1967). *Design of active-site-directed irreversible
   enzyme inhibitors.* Wiley, New York.
3. Hitchings, G. and Burchall, J.J. (1965). *Adv. Enzymol.* **27**, 477.
4. Bertino, J.R., ed. (1971). *Folate antagonists as chemo-therapeutic
   agents. Ann. N.Y. Acad. Sci.* **186**.
5. Dann, J.G., Ostler, G., Bjur, R.A., King, R.W., Scudder, P.,
   Turner, P.C., Roberts, G.C.K., Burgen, A.S.V. and Harding, N.G.L.
   (1976). *Biochem. J.* **157**, 559.

6. Birdsall, B., Griffiths, D.V., Roberts, G.C.K., Feeney, J. and Burgen, A.S.V. (1977). *Proc. Roy. Soc. Lond. (B)* **196**, 251.
7. Feeney, J., Roberts, G.C.K., Birdsall, B., Griffiths, D.V., King, R.W., Scudder, P., and Burgen, A.S.V. (1977). *Proc. Roy. Soc. Lond. (B)* **196**, 267.
8. Kimber, B.J., Griffiths, D.V., Birdsall, B., King, R.W., Scudder, P., Feeney, J., Roberts, G.C.K. and Burgen, A.S.V. (1977). *Biochem.* (in press).
9. Hull, W.E. and Sykes, B.D. (1976). *Biochem.* **15**, 1535.

DIHYDROFOLATE REDUCTASE II:
INTERACTION WITH THE COENZYME, NADPH

J. FEENEY, B. BIRDSALL, G.C.K. ROBERTS
and A.S.V. BURGEN

*Division of Molecular Pharmacology,
National Institute for Medical Research, Mill Hill, London NW7 1AA*

Introduction

NADPH (I) is the coenzyme of dihydrofolate reductase used in the catalysed reduction of dihydrofolate to tetra-hydrofolate. As part of our wider study of ligand binding to the enzyme [1-5] we have been using high resolution NMR spectroscopy to investigate the complexes of NADPH and NADP$^+$ with the enzyme.

The questions we are trying to answer are:

(i) Which are the groups or fragments of the coenzyme which are important in making large contributions to the overall binding?

(ii) What are the differences in the modes of binding of NADPH and NADP$^+$ to the enzyme and can we explain why the reduced form binds a factor of 10^3 more tightly?

(iii) What are the ionisation states of the phosphate groups in the complexes and are they involved directly in the binding?

(iv) Are there any conformational changes in the coenzyme or the enzyme when the complex is formed?

(v) Is it possible to estimate the lifetime of the complexes and thus the dissociation rate constants?

Binding Studies

To investigate which groups on the coenzyme are important for strong binding to the enzyme we have used NMR and fluorescence techniques [6] to measure the binding constants for the various fragments of the coenzyme listed in Table 1.

TABLE 1

Binding constants for NADPH coenzyme analogues binding to L. casei *dihydrofolate reductase*

Ligand	Binding Constant $K_a(M^{-1})$ at pH 7.0
NADPH	1.0×10^8
NADP	4.5×10^4
PADPR	$\sim 10^5$
2',5'-ADP	1.6×10^4
2'-AMP	$2.5 \times 10^3*$
NAD	$<10^2*$
5'-AMP	$>10^2*$

*Determined by NMR methods

The binding constants are seen to increase progressively as we extend the size of the fragments in the series 2'-AMP; 2',5'-ADP; PADPR (which is the fragment lacking the nicotinamide ring); NADPH. A comparison of the binding constants for PADPR and NADPH indicates that the reduced nicotinamide ring is making an important contribution to

the overall binding. However, we have been unable to detect any binding of reduced nicotinamide mononucleotide to
the enzyme. This suggests that when NADPH binds to the
enzyme a conformational change must be induced in the protein to allow the reduced nicotinamide ring to occupy its
favourable binding site. Reduced nicotinamide mononucleotide also shows no detectable binding to the enzyme in the
presence of 2'-AMP which indicates that a larger fragment
of the coenzyme is required to induce the necessary conformation change for optimum binding of the reduced nicotinamide ring.

Coenzyme fragments which do not possess the 2'-phos-
phate group are seen to bind very weakly ($NADP^+$ binds at
least 10^3 times more tightly than NAD^+). We can conclude
from this survey of the binding constants that both the
reduced nicotinamide ring and the 2'-phosphate group play
an important part in the coenzyme binding.

The Role of the 2'-phosphate group in the Binding

The binding constants for ligands such as 2'-AMP can
be conveniently measured by monitoring the ^{31}P chemical
shift of the 2'-phosphate group in 2'-AMP at various concentrations (0.7 to 10.0 mM) in the presence of the enzyme
(1 mM): the bound and free forms of the ligand are in rapid exchange and give rise to a single ^{31}P resonance with a
chemical shift which varies as a function of the 2'-AMP
concentration (see Fig. 1). The observed chemical shift
δ_{obs} is described by the equation

$$\delta_{obs} = \frac{[EL]}{L_T} (\delta_B - \delta_F) + \delta_F \qquad (1)$$

where δ_B and δ_F are the chemical shifts of the bound and
free states respectively, L is the total ligand concentration and [EL] is the concentration of the complex. It
is possible to express [EL] as a function of the total
ligand concentration, total enzyme concentration and the

J. FEENEY ET AL.

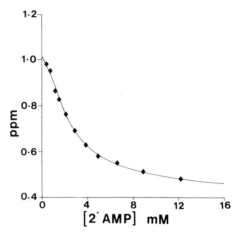

Fig. 1. The ^{31}P chemical shifts of 2'-AMP as a function of increasing
ligand concentration in the presence of 1 mM solution of dihydrofolate reductase at pH 7.0. The ^{31}P shifts were referenced externally
to a 50 mM K$_2$HPO$_4$ solution (pH 8.0) contained in a capillary. Positive shifts to low field.

binding constant K_a as obtained from the mass action equation. By curve fitting the ^{31}P chemical shift data to
equation (1) using a non-linear least squares method one
can obtain values for δ_B and K_a. Experiments of this type
have been carried out for 2'-AMP over a pH range of 4.5
to 7.5 and the results are summarised in Fig. 2. Curves A
and B show the pH dependence of the ^{31}P chemical shifts
of the bound and free 2'-AMP: it is clear from the large
variation in the bound shift with pH that the 2' AMP is
binding in both the dianionic and monoanionic states and
that the pK of the 2'-phosphate group has been reduced on
binding to the enzyme. The Dixon plot shown in Curve C
indicates that the binding constant is substantially decreased as one lowers the pH value from 7.5 to 4.5. A
combined analysis of the data on Curves A and C yields a
value for the pK of bound 2'-AMP of 4.8 ± 0.2 compared
with a value of 6.0 ± 0.05 for free 2'-AMP under these
conditions. This implies that the dianionic form of 2'-AMP
binds 16 fold more tightly to the enzyme than does the

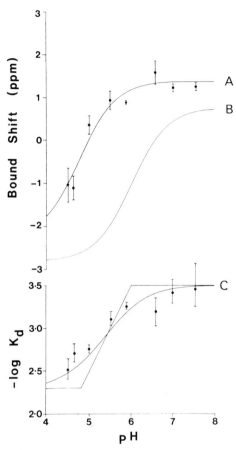

Fig. 2. A. The ^{31}P chemical shifts of the bound 2'-AMP as a function of pH. B. The ^{31}P chemical shifts of free 2'-AMP as a function of pH. C. A Dixon plot of - log K_a for 2'-AMP binding as a function of pH.

monoanionic form. The curves A and C are theoretical curves calculated with these pK values and they are seen to give a good fit to the experimental data. It appears that there is no change in pK values of groups on the en-zyme involved in the binding titrating in the pH range 5–7 when 2'-AMP binds. The ^{31}P shift in the bound dianio-nic form is substantially to lower field (0.49 ppm) of that in the free dianionic form and it seems likely that

this is a consequence of small distortions in the O—P—O
bond angles (with consequent changes in hybridisation)
when the dianionic phosphate binds to the cationic binding
sites on the enzyme. We have undertaken similar experiments with 2',5'-ADP measuring ^{31}P chemical shifts as a
function of ligand concentration in the presence of 1 mM
solution of the enzyme at pH 7.0 (see Fig. 3). In this

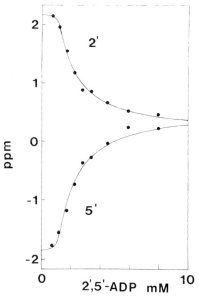

Fig. 3. The ^{31}P chemical shifts of the phosphorus nuclei of 2',5'-ADP
measured as a function of increasing ligand concentration in the
presence of a 1 mM solution of dihydrofolate reductase at pH 7.0.

case we have two ^{31}P resonances which can be readily assigned to the 2'-phosphate (doublet) and 5'-phosphate
(triplet) on the basis of their multiplicities in single
resonance spectra of samples containing a high ratio of
ligand to enzyme. At lower ligand concentrations where
nearly all the ligand is bound to the enzyme the 2'-phos-
phate signal has a chemical shift close to that of the
bound 2'-AMP while the 5'-phosphate has a shift more similar to that of the monoanionic species. This indicates
that the 2'-phosphate is binding in the dianionic form and

the 5'-phosphate is mainly in the monoanionic form (even at this high pH value).

When we examine the ^{31}P spectra of the tightly bound coenzyme NADPH in the presence of the enzyme we have slow exchange conditions on the NMR timescale and separate ^{31}P signals are observed for the free and bound ligands (4). The ^{1}H noise decoupled ^{31}P spectrum of NADPH (see Fig. 4A) consists of a low field signal for the 2'-phosphate group

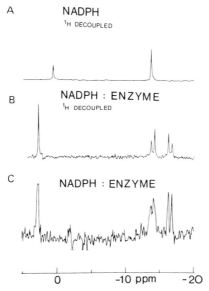

Fig. 4. A. The ^{31}P proton noise-decoupled spectrum of 10 mM NADPH at pH 6.9 at 40.5 Hz. B. The ^{31}P proton noise-decoupled spectrum of 1.3 mM NADPH in the presence of 1.3 mM dihydrofolate reductase at pH 6.9. C. The ^{31}P single resonance spectrum of 1.1 mM NADPH in the presence of 1.1 mM dihydrofolate reductase at pH 6.9.

a high field signal with twice the intensity corresponding to the two pyrophosphate nuclei which accidentally have the same chemical shift. When an equimolar solution of NADPH (1.3 mM) and dihydrofolate reductase (1.3 mM) is examined, the 1:1 stoichiometric complex thus formed shows dramatic spectral changes (Fig. 4B). The two pyrophosphate ^{31}P nuclei are shifted to high fields (-13.94 and -16.47 ppm) by different amounts (the ^{31}P–^{31}P spin coupling is

now manifested as doublets on the two non equivalent resonances) and the 2'-phosphate resonance is shifted to low
field (2.66 ppm) near to the shift position of bound dia-
nonic phosphate in the fragments 2'-AMP and 2',5'-ADP.
Similar experiments with NADP[+] yielded almost exactly the
same shift (2.72 ppm) for the 2'-phosphate and similar
shifts for the pyrophosphate (-14.32 and -16.23 ppm). The
2'-phosphate groups in the free coenzymes have pK values
of 6.1 for NADPH and 6.4 for NADP[+] and the ^{31}P resonances
of the 2'-phosphate groups show a corresponding change in
chemical shift with pH (see Fig. 5). However, for the

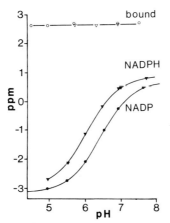

Fig. 5. Variation of the ^{31}P chemical shift of the 2'-phosphate resonance of NADP[+] (o,o) and NADPH (∇,∇) with pH. The open symbols refer
to the coenzyme bound to dihydrofolate reductase and the solid symbols
to free coenzymes. All pH values are uncorrected meter readings (glass
electrode) from D$_2$O solutions.

coenzyme-enzyme complexes the 2'-phosphate resonance does
not titrate within the range pH 4.5 to 7.5 (see Fig. 5)
indicating that the pK of the bound species has been lowered by at least 3 pK units compared with the value in the
free coenzyme. This fact, together with the evidence for
O–P–O angle distortion on binding, strongly suggests that
the dianionic 2'-phosphate group is involved in a strong
electrostatic interaction with cationic sites on the
enzyme.

Comparison of the complexes of NADP$^+$ and NADPH with the enzyme

The similar shifts observed for the 2'-phosphate ^{31}P signals in the complexes for NADP$^+$ and NADPH indicate that the 2'-phosphate groups in both forms of the coenzyme are binding in the same ionisation state and at the same binding site in the two complexes. Furthermore the similarity of the pyrophosphate ^{31}P shifts for the two complexes also shows that this fragment of the molecules also binds in a similar, though not identical, manner in the two complexes. A consideration of the ^{1}H shifts of the A2 protons in the bound coenzyme provides evidence that the adenine rings are also binding in the same binding sites in the complexes. By examining the ^{1}H spectra of equimolar solutions of coenzyme and the enzyme (selectively deuterated as discussed in the previous paper) one can easily detect the A2 proton for the bound coenzymes and assign the signals unequivocally by transfer of saturation experiments (Fig. 6). In each case the resonance of the A2 proton has been shifted by 0.9 ppm to high field of its position in the free coenzyme clearly showing the adenine ring to be binding in the same environment in the two complexes.

It can be concluded from these findings that the differences in binding between the NADP$^+$ and NADPH are strongly localised at the binding site for the nicotinamide ring. The nicotinamide rings in the oxidised and reduced forms are very similar in size and shape and would be expected to bind fairly similarly if they had equal access to the favourable binding site. A plausible explanation for the reduced binding of NADP$^+$ is that the positive charge of the oxidised nicotinamide ring might interact repulsively with a similar charge on the enzyme and thus cause the nicotinamide ring to be displaced away from the binding site used by the reduced nicotinamide ring.

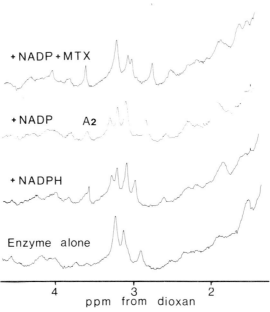

Fig. 6. The aromatic region of the 270 MHz ^{1}H spectra of selectively deuterated dihydrofolate reductase alone, in its complex with NADPH, NADP$^+$ and its ternary complex with NADP$^+$ and methotrexate (MTX).

Lifetime of the NADP-Enzyme Complex

The ^{1}H resonance signal of the A2 proton of NADP$^+$ bound to the enzyme (Fig. 6A) is considerably broader than that from the same proton in the NADPH complex (Fig. 6B). The additional line broadening is due to an exchange contribution to the linewidth of the signal which is not fully in the slow exchange conditions. Addition of one equivalent of methotrexate to the NADP-enzyme complex sharpens the A2 resonance (Fig. 6C) as would be expected if methotrexate increases the binding of NADP$^+$ by reducing the dissociation rate constant. We have measured similar effects in the ^{13}C spectra of carboxamide-^{13}C labelled NADP bound to the enzyme (Fig. 7A) [5]. The enrichment (90 per cent) ensures that there is no problem in distinguishing the ligand carbonyl carbon from those in the protein (natural abundance 1.11 per cent). When three equivalents

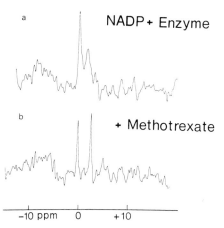

Fig. 7. The ^{13}C NMR spectra at 25.2 MHz of (A) [^{13}CO]NADP$^+$ in the presence of dihydrofolate reductase (molar ratio 3:1) (B) the [^{13}CO]NADP$^+$-dihydrofolate reductase-methotrexate complex. The sample of [^{13}CO]NADP$^+$ was prepared by J.L. Way [5].

of the labelled NADP are added to the 1 mM enzyme solution two carbonyl resonances (intensities 2:1) are observed: the larger signal resonates at the position of free NADP and the other, corresponding to bound NADP, resonates at 1.6 ppm to higher field of this value. Clearly these broad signals are approaching the slow exchange condition and are broadened by an exchange contribution. Addition of methotrexate removes the line broadening (Fig. 7b) and allows one to estimate the line-widths in the absence of exchange. A full line shape analysis of the broadened spectrum shown in Fig. 7a yields a value for the dissociation rate constant of 45 ± 10 sec^{-1}. The equilibrium constant for the formation of the binary NADP$^+$-enzyme complex is $\approx 4.5 \times 10^4$ M^{-1} and combining this with the K_{-1} calculated above gives a K_1 value of $\sim 2 \times 10^6$ M^{-1} sec^{-1} which is only slightly less than the association rate constant expected from a diffusion controlled reaction.

Conformational Changes on Complex Formation

In the previous paper [7] we presented evidence that the ^{19}F resonance of one of the tyrosines on the 3-fluoro-tyrosine labelled enzyme was shifted upfield as a result of a conformational change when NADPH binds to the enzyme-methotrexate complex [8]. Examination of the ^{1}H spectra of the enzyme-coenzyme complexes also reveal several spectral changes on complex formation. For example, three of the tyrosine resonances are influenced by coenzyme binding and one of them which is shifted in opposite directions by the oxidised and reduced forms of the coenzyme remains un-shifted when the fragment lacking the nicotinamide ring binds to the enzyme. However, at this stage we have not been able to distinguish which of these changes are caused by direct effects of the ligand binding and which result from indirect effects associated with induced conforma-tional changes.

When we consider the conformational state of the bound coenzyme we can be more definitive in our conclusions. In the ^{31}P spectrum of the NADPH-enzyme complex we can detect the effects of phosphorus proton coupling constants on the pyrophosphate ^{31}P resonances by observing the spectrum in the absence of proton noise decoupling (Fig. 4C). Three-bond coupling of the phosphorus to the 5'CH$_2$ protons is known to be conformationally sensitive to the P–O–C–H di-hedral angle (J_{gauche} = 1.8 Hz, J_{trans} = 20.6 Hz) [9]. We have simulated the experimental spectrum shown in Fig. 4C and estimate that for the low field pyrophosphate sig-nal the sum of the coupling constants to the 5'CH$_2$ protons is at least 13 Hz and substantially less ($<$ 5 Hz) on the high field pyrophosphate. (The actual coupling constants are larger than the observed line broadenings because of rapid proton relaxation influencing the shape of the mul-tiplets.) The phosphorus nucleus giving rise to the high field signal is thus consistent only with a conformation in which it is gauche to both 5' protons. This is the conformation which is favoured for 5'-nucleotides [10] and NADPH [11] in solution. The other pyrophosphate nucleus shows couplings which indicate it to be almost eclipsed with one of the 5' protons.

Conclusions

In these studies we have been able to show that in the coenzyme binding to dihydrofolate reductase:

(i) The 2'-phosphate group and the reduced nicotinamide ring make important contributions to the overall binding energy.

(ii) The 2'-phosphate group binds in its dianionic state and is involved in strong electrostatic interactions with cationic group(s) on the enzyme.

(iii) The adenine ring, the ribose 5'-phosphate and (to a lesser extent) the pyrophosphate fragments of NADP$^+$ and NADPH bind in a similar manner to the enzyme and the main difference in binding is localised at the nicotinamide ring binding site.

(iv) There is a specific conformational change in one of the pyrophosphates of NADPH when it binds to the enzyme involving a 50° change in the P—O—C—H torsion angle.

References

1. Roberts, G.C.K., Feeney, J., Burgen, A.S.V., Yuferov, V., Dann, J.G. and Bjur, R.A. (1974). *Biochemistry* **13**, 5351.
2. Birdsall, B., Griffiths, D.V., Roberts, G.C.K., Feeney, J. and Burgen, A.S.V. (1977). *Proc. Roy. Soc.* **196**, 251.
3. Feeney, J., Roberts, G.C.K., Birdsall, B., Griffiths, D.V., King, R.W., Scudder, P. and Burgen, A.S.V. (1977). *Proc. Roy. Soc.* **196**, 267.
4. Feeney, J., Birdsall, B., Roberts, G.C.K., Burgen, A.S.V. (1975). *Nature* **257**, 564.
5. Way, J.L., Birdsall, B., Feeney, J., Roberts, G.C.K. and Burgen, A.S.V. (1975). *Biochemistry* **14**, 3470.
6. Dunn, S. unpublished results.
7. Roberts, G.C.K., Feeney, J., Birdsall, B., Kimber, B.J., Griffiths, D.V., King, R.W. and Burgen, A.S.V. Previous paper this volume.
8. Kimber, B.J., Griffiths, D.V., Birdsall, B., King, R.W., Scudder, P., Feeney, J., Roberts, G.C.K. and Burgen, A.S.V. (1977). Submitted for publication.

9. Blackburn, B.J., Lapper, R.D. and Smith, I.C.P. (1973). *J. Amer. Chem. Soc.* **95**, 2873.
10. Davies, R.B. and Danyluk, S.S. (1974). *Biochemistry* **13**, 4417.
11. Sarma, R.H. and Mynott, R.J. (1973). *In: Conformation of Biological Molecules and Polymers*, eds. Bergmann, E.D. and Pullman, B. 591 (Israel Academy of Science and Humanities, Jerusalem).

THE ARCHITECTURE OF
AN ANTIBODY COMBINING SITE

PETER GETTINS and RAYMOND A. DWEK

Department of Biochemistry, Oxford

Introduction

Antibodies are the basis of the body's defence. The molecular basis of antibody-antigen recognition is therefore of fundamental importance. The essence of the immune system involves diversity in recognition and therefore a comparative analysis of a large number of combining sites is required to provide the details of molecular recognition by antibodies. It is also important to develop new methodologies by which the antibody combining site can be analysed in solution and yield information comparable with that obtained from crystal studies. We describe here an attempt to elucidate the structure of the binding site of the Dnp-binding mouse IgA myeloma protein, 315, by a combination of high resolution NMR, ESR, model building and chemical modification. We show that by this approach it is possible to determine the general dimensions of the site, its polarity and asymmetry features and to assign the contact residues with the ligand as well as their three-dimensional coordinates. This approach, when applied to other antibodies, will then provide information about the structural basis of antibody specificity.

Basic Antibody Structure

From chemical and enzymic studies [1,2] it has been
possible to establish the multiple chain structure of
antibody molecules and to attribute the distinct functions
of antigen binding and complement fixation to separate
regions of the antigen binding and complement fixation to
separate regions of the molecule (Fig. 1). The question of
how such a closely related series of proteins as the immunoglobulins can have such a wide range of specificities
can only be answered if it is known exactly what gives
rise to specificity in a given case. To answer this it is
necessary to know which amino-acid residues are in contact with the hapten and their relative positions. Some
information has come from affinity labelling of residues
close to the combining site. In all cases the side chains
labelled have been in the regions of maximum sequence
variability, called hypervariable regions. It is these
regions that contain the residues responsible for antibody specificity and diversity.

Crystal structure provides more detailed information
about the combining site. However, for such studies homogeneous antibodies are required, which cannot be obtained
from normal serum due to the highly heterogeneous population in an immune serum. However, certain cancers produce
myeloma proteins which are homogeneous and chemically indistinguishable from normally raised antibodies. The
structures of several myeloma proteins have now been determined [3,4,5,6]. Of particular importance to the work
discussed here are the structures of the antigen binding
fragments (Fabs) of proteins New [3] and 603 [4], which
bind vitamin K_1OH and phosphorylcholine respectively.
From these structures it was seen that there are distinct
domains, comprising 110–120 residues folded in two layers
of anti-parallel β-pleated sheet and connected by a disulphide bridge. This unit has been termed the *immunoglobulin fold* (Fig. 2). One domain from each chain then asso-

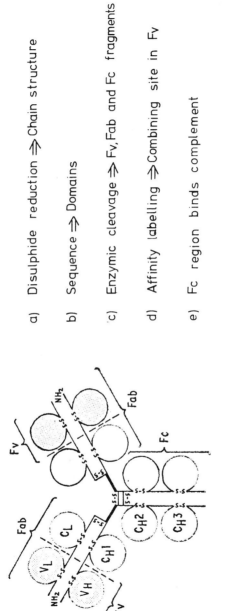

a) Disulphide reduction ⇒ Chain structure

b) Sequence ⇒ Domains

c) Enzymic cleavage ⇒ Fv, Fab and Fc fragments

d) Affinity labelling ⇒ Combining site in Fv

e) Fc region binds complement

Fig. 1. The structural features of immunoglobulins, as deduced from chemical studies.

ciates to form the larger globular units. Two of these
form an Fab region while one constitutes the region Fv.
The Fv region contains the hypervariable residues and the
importance of these in the recognition process has been
shown from the crystal structures of proteins 603 and New
when bound to their respective haptens. These hypervariable residues occur together in space as loops, between
strands of the immunoglobulin fold and provide all the
hapten contact residues.

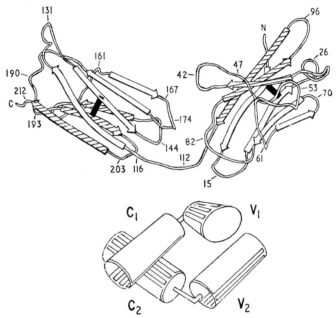

Fig. 2. Top: The structure of one chain of the human light chain dimer Mcg; bottom: The mode of association of the two chains to form
the dimer.

Structure Prediction of Antibody Combining Sites

The division of the sequence in the variable region
into those residues which serve a functional role and
those which are purely structural, provides a basis for
predicting the combining site geometry of other antibodies
whose sequences are known, by building the hypervariable
loops onto the immunoglobulin fold of a determined struc-

ture. To be able to do this with any degree of certainty
there should be a significant amount of homology between
the sequences of the known structure and the one to be
predicted. The changes in residues will then be concentrated in the hypervariable regions, where, it is assumed,
there will be only minor perturbations of the fold region.
The two sequences are aligned so that there is maximum
homology between them, which may mean leaving gaps between
residues in one chain to allow for an insertion in the
other (Table 1). This then shows which residues correspond
to the hypervariable regions and these can then be built
onto the immunoglobulin fold framework. The conformation
of the loops is based on the original structure where
possible, if this does not lead to hydrophilic residues
in a hydrophobic environment or vice versa.

For protein 315, a Dnp-binding mouse myeloma immunoglobulin, the structures of proteins 603 and New provide
an excellent basis for such a prediction since there is
40—50 per cent homology in sequence between them. Padlan
et al. [7] have described the model of the combining site
of protein 315 built using the coordinates of these two
immunoglobulin fragments (Fig. 3). Its main feature is a
concentration of aromatic side chains surrounding the combining site, which is a cleft 1.2 nm deep between the
heavy and light chains. By using the data from ESR and NMR
experiments presented here it is possible to obtain structural information about the residues directly probed by
the various haptens and thus to obtain a detailed picture
of the combining site by refinement of the original predicted model.

Combining Site Dimensions

Both the depth of the combining site and its lateral
dimensions at the entrance were determined using a series
of Dnp haptens with six membered nitroxide rings linked
to the Dnp ring by a spacer group. The method compares the

TABLE 1A

Alignment of V_L Sequences

MOPC315	1	Pca-Ala-Val-Val-Thr-Glu-Glu- -Ser-Ala-Leu-Thr-Thr-Ser-Pro-Gly-Gly-Thr-Val-Ile-Leu-Thr-Cys-Arg-Ser-
McPC603	1	Asp-Ile-Val-Met-Thr-Gln-Ser-Pro-Ser-Ser-Leu-Ser-Val-Ser-Ala-Gly-Glu-Arg-Val-Thr-Met-Ser-Cys-Lys-Ser-
New	1	Pca-Ser-Val-Leu-Thr-Gln-Pro-Pro- -Ser-Val-Ser-Gly-Ala-Pro-Gly-Gln-Arg-Val-Thr-Ile-Ser-Cys-Thr-Gly-
Mcg	1	Pca-Ser-Ala-Leu-Thr-Gln-Pro-Pro- -Ser-Ala-Ser-Gly-Ser-Leu-Gly-Gln-Ser-Val-Thr-Ile-Ser-Cys-Thr-Gly-
Rei	1	Asp-Ile-Gln-Met-Thr-Gln-Ser-Pro-Ser-Ser-Leu-Ser-Ala-Ser-Val-Gly-Asp-Arg-Val-Thr-Ile-Thr-Cys-Gln-Ala-

MOPC315	25	Ser-Thr-Gly-Ala-Val- -Thr-Thr-Ser-Asn-Tyr-Ala-Asn-Trp-Ile-Glx-Glx-Lys-Pro-Asx-His-Leu-Phe-
McPC603	26	Ser-Glx-Ser-Leu-Leu-Asx-Ser-Gly-Asx-Glx-Lys-Asx-Phe-Leu-Ala-Trp-Tyr Glx-Glx-Lys-Pro-Gly-Glx-Pro-Pro-
New	25	Ser-Ser-Asn-Ile- -Gly-Ala-Gly-Asn-His-Val-Lys-Trp-Tyr-Gln-Gln-Leu-Pro-Gly-Thr-Ala-Pro-
Mcg	25	Thr-Ser-Asp-Val- -Gly-Gly-Tyr-Asn-Tyr-Val-Ser-Trp-Tyr-Gln-Gln-His-Ala-Gly-Lys-Ala-Pro-
Rei	26	Ser-Gln- -Asp-Ile-Lys-Tyr-Leu-Asn-Trp-Tyr-Gln-Gln-Thr-Pro-Gly-Lys-Ala-Pro-

MOPC315	45	Thr-Gly-Leu-Ile-Gly-Gly-Thr-Ser-Asp-Arg-Ala-Pro-Gly-Val-Pro-Val-Arg-Phe-Ser-Gly-Ser-Leu-Ile-Gly-Asp-
McPC603	45	Lys-Leu-Leu-Ile-Tyr- X - X - X - X - X - X -Gly-Val-Pro-Ala-Arg-Phe-Ser-Gly-Ser-Gly-Ser-Arg-Thr-
New	45	Lys-Leu-Leu-Ile-Phe-His-Asn-Asn-Ala-Arg- -Phe-Ser-Val-Ser-Lys-Ser-Gly-Ser-
Mcg	45	Lys-Val-Ile-Ile-Tyr-Gly-Val-Asn-Lys-Arg-Pro-Ser-Gly-Val-Pro-Asp-Arg-Phe-Ser-Gly-Ser-Lys-Ser-Gly-Asn-
Rei	45	Lys-Leu-Leu-Ile-Tyr-Glu-Ala-Ser-Asn-Leu-Gln-Ala-Gly-Val-Pro-Ser-Arg-Phe-Ser-Gly-Ser-Gly-Ser-Gly-Thr-

MOPC315	70	Lys-Ala-Ala-Leu-Thr-Ile-Thr-Gly-Ala-Glx-Thr-Gly-Ala-Met-Tyr-Phe-Cys-Ala-Leu-Trp-Phe-Arg-Asx-
McPC603	70	Asp-Phe-Thr-Leu-Thr-Ile-Asx-Pro-Val-Glx-Ala-Asx-Val-Ala-Thr-Thr-Phe-Cys- X - X - X - X - X -
New	70	Ser-Ala-Thr-Leu-Ala-Ile-Thr-Gly-Leu-Gln-Ala-Glu-Asp-Glu-Ala-Asp-Tyr-Tyr-Cys-Gln-Ser-Tyr-Asp-Arg-Ser-
Mcg	70	Thr-Ala-Ser-Leu-Thr-Val-Ser-Gly-Leu-Gln-Ala-Glu-Asp-Glu-Ala-Asp-Tyr-Tyr-Cys-Ser-Ser-Tyr-Glu-Gly-Ser-
Rei	70	Asp-Tyr-Thr-Phe-Thr-Ile-Ser-Ser-Leu-Gln-Pro-Glu-Asp-Ile-Ala-Thr-Tyr-Tyr-Cys-Gln-Gln-Tyr-Gln-Ser-Leu-

MOPC315	97	-His-Phe-Val-Phe-Gly-Gly-Gly-Thr-Lys-Val-Thr-Val-Leu-Gly.
McPC603	97	- X - X - X -Phe-Gly-Gly-Gly-Thr-Lys-Leu-Thr-Val-Leu-Lys-Arg.
New	94	-Leu-Arg-Val-Phe-Gly-Gly-Gly-Thr-Lys-Leu-Thr-Val-Leu-Arg.
Mcg	95	Asp-Asn-Phe-Val-Phe-Gly-Thr-Gly-Thr-Lys-Val-Thr-Val-Leu-Gly.
Rei	95	-Pro-Tyr-Thr-Phe-Gly-Gln-Gly-Thr-Lys-Leu-Gln-Ile-Thr-Gly.

TABLE 1B

Alignment of V$_H$ Sequences

MOPC315 1 Asp-Val-Gln-Leu-Gln-Glu-Ser-Gly-Pro-Gly-Leu-Val-Lys-Pro-Ser-Gln-Ser-Leu-Ser-Leu-Thr-Cys-Ser-Val-Thr-
McPC603 1 Glu-Val-Lys-Leu-Val-Glu-Ser-Gly-Gly-Gly-Leu-Val-Gln-Pro-Gly-Gly-Ser-Leu-Arg-Leu-Ser-Cys-Ala-Thr-Ser-
New   1 Pca-Val-Glu-Leu-Pro-Glu-Gly-Pro-Glu-Leu-Val-Ser-Pro-Arg-Glx-Thr-Leu-Ser-Leu-Thr-Cys-Thr-Leu-Ser-

MOPC315 26 Gly-Tyr-Ser-Ile-Thr-Ser-Gly-Tyr-Phe-Trp-Asn-Trp-Ile-Arg-Gln-Phe-Pro-Gly-Asn-Lys-Leu-Glu-Trp-Leu-Gly-
McPC603 26 Gly-Phe-Tyr-Phe-Ser-Asp-Phe-Tyr-Met-   -Glu-Trp-Val-Arg-Gln-Pro-Pro-Gly-Lys-Arg-Leu-Glu-Trp-Ile-Ala-
New   26 Gly-Ser-Ser-Val-Phe-Ala-Val-Tyr-Ile-   -Val-Trp-Val-Arg-Gln-Pro-Pro-Gly-Arg-Gly-Leu-Glu-Trp-Ile-Ala-

MOPC315 50 Phe-Ile-Lys-Tyr-Asp-Gly-   -Ser-Asx-Tyr-Gly-Asx-Pro-Ser-Leu-Lys-Asn-Arg-Val-Ser-Ile-Thr-Arg-
McPC603 50 Ala-Ser-Arg-Asx-Lys-Gly-Asn-Lys-Tyr-Thr-Thr-Glu-Tyr-Ser-Ala-Ser-Val-Lys-Gly-Arg-Phe-Ile-Val-Ser-Arg-
New   50 Tyr-Val-Phe-Tyr-His-Gly-   -Thr-Ser-Asp-Thr-Pro-Leu-Arg-Leu-Arg-Ser-Arg-Val-Thr-Met-Leu-Val-

MOPC315 72 Asp-Thr-Ser-Glu-Asn-Gln-Phe-Phe-Leu-Lys-Leu-Asp-Ser-Val-Thr-Thr-Glx-Asx-Thr-Ala-Thr-Tyr-Tyr-Cys-Ala-
McPC603 72 Asp-Thr-Ser-Glu-Ser-Ile-Leu-Tyr-Leu-Gln-Met-Asn-Ala-Leu-Arg-Ala-Glu-Asp-Thr-Ala-Ile-Tyr-Tyr-Cys-Ala-
New   72 Asn-Thr-Ser-Lys-Asn-Glu-Phe-Ser-Leu-Arg-Leu-Ser-Val-Thr-Ala-Ala-Asp-Thr-Ala-Val-Tyr-Tyr-Cys-Ala-

MOPC315 98 Gly-Asp-Asn-Asp-His-   -Leu-Tyr-Phe-Asp-Tyr-Trp-Gly-Gln-Gly-Thr-Thr-Leu-Thr-Val-Ser-Ser.
McPC603 97 Arg-Asn-Tyr-Tyr-Gly-Ser-Thr-Trp-Tyr-Phe-Asp-Val-Trp-Gly-Ala-Gly-Thr-Thr-Val-Thr-Val-Ser-Ser.
New   97 Arg-Asx-Leu-Ile-Ala-   -Gly-Cys-Ile-Asx-Val-Trp-Gly-Gln-Gly-Ser-Leu-Val-Thr-Val-Ser-Ser.

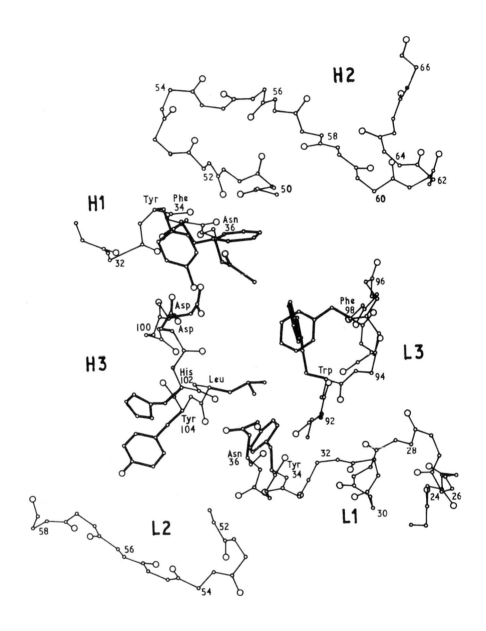

Fig. 3. The original predicted hypervariable loops of protein 315. H1, H2 and H3 are from the heavy chain and L1, L2 and L3 are from the light chain.

observed ESR hyperfine splittings of the nitroxide with
those calculated on the basis of different postulated
motions for the spin labels. Correlation of these motions
with the inherent mobilities of the haptens, derived from
molecular models, then gives the constraints imposed by
the combining site and thus its dimensions. These dimensions of 1.1—1.2 nm deep and 0.6 × 0.9 nm wide are in good
agreement [8] with those obtained from the predicted model
of the combining site [7]. Extension of this method to
five membered nitroxide spin labels allows discrimination
between the two sides of the combining site, since each
enantiomer gives rise to a separate bound signal (Fig. 4).
This indicates that the combining site is asymmetric [8]
one side of which is narrow enough to hold the spin-label
ring rigidly, while the other allows some motion (Fig. 5).

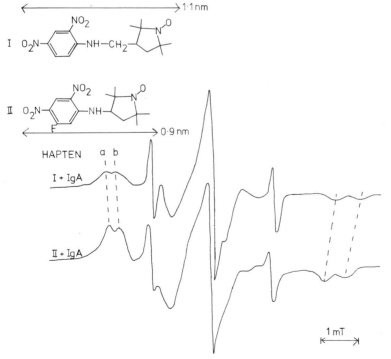

Fig. 4. The ESR spectra of five membered nitroxide spin labels bound
to protein 315, showing the two bound signals (a and b) arising from
the two enantiomers of each hapten.

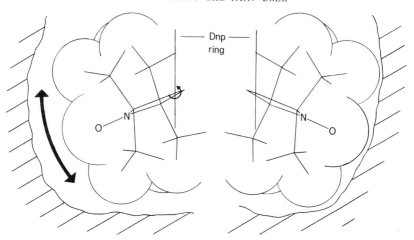

Fig. 5. Two enantiomers of Hapten II bound to protein 315. The arrows indicate the bond about which rotation is possible, due to the greater width of the left side of the combining site, and the extent of motion allowed (40°).

Table 2 summarises the results for five membered spin labels.

Assignment of Histidine Residues

By virtue of the large chemical shift separation of the protonated and unprotonated forms, it is very easy to monitor the resonances of histidines by NMR (Fig. 6). As the Fv fragment of protein 315 contains 3 histidine residues – His 44_L, His 97_L and His 102_H – their assignment from the model to particular NMR resonances provides convenient internal markers to monitor perturbations in the protein. His 44_L, 2.0 nm from the combining site, was assigned from spin label difference spectroscopy, in which resonances from His 44_L and His 97_L, which are close to the combining site, are broadened out [9]. His 97_L and His 102_H are distinguished by their pK_as, line widths and chemical shifts. The full assignment, discussed elsewhere [10], is that the residue with pK_a 6.9 is His 102_H, that with pK_a 5.9 is His 97_L and that with pK_a 8.2 is His 44_L [10]. The predicted model shows that the position of His 102_H is at

TABLE 2

Results from ESR Studies of Five Membered Spin Labelled Haptens

Hapten	Calculated $A_{z'z'}$ for motion (mT)	Observed $A_{z'z'}$* (mT)		Requirements for motion (nm)		
				Width	Height	Depth
V Dmp-NH-CH$_2$	2.81	3.50	2.85	0.7	1.0	1.0
VI 5-F-Dnp-NH	3.10	3.34	2.65	0.6	0.9	0.8

*The observed values of $A_{z'z'}$ have been corrected for the effects of molecular tumbling.

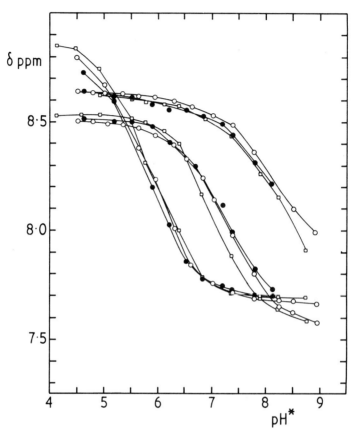

Fig. 6. ^{1}H 270 MHz NMR titration curves for the three histidines in Fv.

the entrance of the combining site and should therefore be very sensitive to the hapten side chain position. In addition, the area of each histidine resonance is a useful standard from which to estimate the number of protons perturbed on hapten binding and thus appearing in the difference spectra (see later).

Distribution of Charged Residues

In addition to probing different areas of the combining site, the more rigidly held enantiomer of the spin labelled hapten I experiences the effect of an electrostatic field gradient due to a nearby positive charge. This charge increases the contribution from the dipolar form of the nitroxide bond $\overset{+}{N}{-}O^{-}$ and leads to an observed hyperfine ESR splitting $(A_{z'z'})$ value greater than expected solely from constraints on mobility. The $A_{z'z'}$ value of 3.56 mT (corrected for the effects of molecular tumbling) is greater than the rigid limit value of 3.4 mT [8]. This can only be explained by a positive charge in van der Waals contact with the nitroxide group. The distance of the positive charge is calculated from an inverse square distance relationship [8].

From the predicted model it is then possible to assign the positively charged residue interacting with the nitroxide as Arg 95_L. This does however depend on the mode of hapten binding. The main constraint is stacking with the side chain of Trp 93_L. The only mode of binding which is consistent with the hapten mobilities deduced from the ESR data has the 2-NO_2 group hydrogen bonded to Asn 36_L and the 4-NO_2 group pointing towards Phe 34_H, which is different from that presented later for other Dnp haptens and probably results from the extremely bulky side chains of these haptens.

A subtle method for providing more evidence for charged groups at the entrance to the combining site uses ^{31}P NMR studies on related Dnp and Tnp phosphoryl haptens. The proximity of the phosphorus to a charged group is detected by a change in the pK_a of the phosphoryl group. Table 3 shows that for the three closely related Tnp alkyl phosphonates (methyl, ethyl and propyl) only the longest has its pK_a significantly perturbed [10], thus placing a positive charge at the entrance to the combining site in the same position as that deduced for the ESR studies. ^{1}H NMR shows that His 102_H has its pK_a changed by all three haptens, so that it cannot be the perturbing positive charge. Only two other positively charged residues in the site are capable of interacting with hapten side chains: Lys 52_H and

TABLE 3

pK$_a$ values of phosphonate and phosphate groups of phospho-haptens
free and when bound to the Fv fragment from MOPC 315

pK$_a$ values were determined from ^{31}P NMR titrations at 129 MHz. All measurements were made at 293K in the presence of 0.15 M NaCl. The errors are those obtained from analysis of the titration curves. The pK$_a$s of histidine 102$_H$(pK$_a$ 6.9) were obtained from 270 MHz ^{1}H NMR spectra.

Phospho-hapten	pK$_a$ free	pK$_a$ bound	102$_H$ histidine pK$_a$
TNP.NH.CH$_2$.CH$_2$.CH$_2$PO$_3$H$^-$	7.7	6.9 ± 0.2	7.2
TNP.NH.CH$_2$.CH$_2$PO$_3$H$^-$	6.9	7.1 ± 0.2	7.3
TNP.NH.CH$_2$PO$_3$H$^-$	5.7	5.9 ± 0.2	7.3
DNP.NH.CH$_2$PO$_3$H$^-$	6.0	6.1 ± 0.2	7.3
DNP.NH.CH$_2$.CH$_2$OPO$_3$H$^-$	6.3	6.6 ± 0.1	7.4

Arg 95_L. Of these Lys 52_H has been shown by affinity labelling to be close to the Dnp binding site. On these data
alone it is not possible to distinguish between them but
the conclusion that only Arg 95_L can interact with the
nitroxide group of the spin labelled hapten and the similar lengths of this hapten and the Tnp propyl phosphonate
hapten suggests that this residue is again responsible.
Recent chemical modification studies [11] support this interpretation.

The Dnp Binding Site

The properties of the dinitrophenyl ring (having a
large dipole moment and being a good electron acceptor)
make it very likely that the greatest energy of interaction with an amino acid side chain would arise with tryptophan, by forming either a dipole-dipole interaction or a
charge-transfer type complex. This would then give a large
contribution to the observed binding energies for Dnp haptens. Observation of red shifts in the UV absorption spectra of Dnp haptens on binding to protein 315 [12] and circular dichroism studies [12,13] both indicate that a
favourable interaction is indeed occurring between Dnp
and the indole ring of tryptophan.

Solution studies by NMR [15] were used to investigate
the geometry of interaction of Dnp with tryptophan by
relating the shifts experienced by all the aromatic ring
protons to the position of each proton relative to the
centre of the other aromatic ring. This analysis assumes
that the shifts are caused exclusively by ring current induced local magnetic fields [16,17] and ignores perturbations caused by the indole side chain and the two nitro
groups. The two main theories, those of Johnson and Bovey
[16] and Haigh and Mallion [17] are currently used to calculate ring current induced magnetic fields of aromatic
rings. For upfield shifts, arising from protons either
above or below the plane of the ring, it is generally

accepted that the Johnson-Bovey treatment is the better, since that of Haigh and Mallion gives large underestimates [17]. Since only upfield-shifted protons are studied in this work, the Johnson-Bovey theory is used throughout. However it is necessary to extend the basic treatment for six membered rings to five membered rings to take into account the pyrrole ring of tryptophan. Details of this extension are given in the Appendix of ref. [15]. Observation of upfield shifts on all aromatic protons, of comparable magnitudes for the protons of each ring, immediately restricts the geometry of the complex to one in which both rings are parallel or near-parallel and overlapping one another significantly. By assuming an inter-ring separation of 0.33 nm [18] a more precise structure was arrived at (Fig. 7). Although only approximate because of the inherent uncertainties, it is probably accurate to about 0.05 nm.

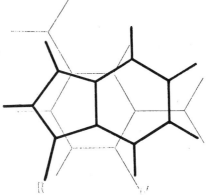

Fig. 7. Structure of Dnp-aspartate/Tryptophan complex in solution. R and R' represent the side chains. The view is perpendicular to the planes of both rings which are parallel and separated by 0.33 nm.

The presence of a tryptophan in the combining site is also indicated by the model of Padlan et al. [7], which has the side chain of Trp 93_L lining one side of the Dnp binding pocket, so that a similar geometry to that obtained for the Dnp/tryptophan complex would be possible in the Fv. For such an interaction, the expected shifts on the three Dnp aromatic protons on binding to protein 315 would be ~0.2--0.6 ppm.

Addition of various haptens to Fv results in large upfield shifts on all three Dnp proton resonances [15], as well as perturbations of aromatic protein resonances which are best displayed as difference spectra (Fig. 8).

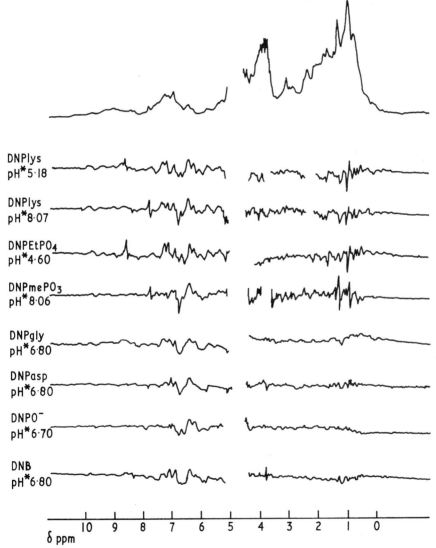

Fig. 8. Difference spectra between Fv in the presence and absence of a variety of haptens at 303 K. Spectra were recorded at 270 MHz.

Since all hapten shifts are greater than 1 ppm and in some
cases greater than 2 ppm, it is impossible to account for
these solely on the basis of an interaction with one aromatic protein side chain. To account for the large hapten
shifts it is necessary to have two additional aromatic
rings close to the Dnp protons. In the case of Dnp-glycine
not only the Dnp protons but also the two α-CH$_2$ protons
experience large shifts (Table 4) so that a total of 4

TABLE 4

Dnp-glycine/Fv Titration: NMR Chemical Shift Changes.
The changes caused by the hapten on the protein are accurate
to 0.05 ppm, those by protein on the hapten to 0.01 ppm

		δo	$\Delta\delta$ ppm
Changes on Hapten	H(3)	9.12	-1.21
	H(5)	8.30	-2.20
	H(6)	6.94	-1.77
	CH$_2$	4.13	-1.00
Changes on Fv		6.8	-0.23
		6.3	+0.33
		6.1	-0.14
		5.5	+0.23

aromatic side chains are needed as contact residues. Fig.
9 shows the resultant configuration. The predicted shifts
by Dnp on the protein for this geometry are in good agreement with observations in the model systems, being small
in magnitude but both up and downfield depending on whether they originate from tryptophan or the perpendicular
aromatics respectively. The model shows that the only
aromatic residues capable of interacting in this way are
Phe 34$_H$, Tyr 34$_L$ and Tyr 33$_H$. These affect H(5) and H(6),
H(3) and the α-CH$_2$ group of Dnp-glycine respectively.

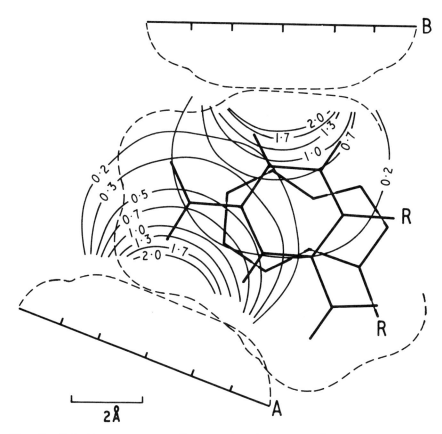

Fig. 9. Relative geometry of aromatic rings around the hapten which is required to account for the upfield shifts on the hapten on binding to Fv. The contours represent lines of equal ring current shift, in ppm upfield.

Since only the equivalent of about 10 aromatic protons are perturbed by hapten binding, the 5 which move upfield can probably be assigned to some of the resonances of Trp 93_L, while some of the downfield shifted resonances are attributed to Tyr 34_L and Phe 34_H.

In order for these aromatic rings to have the required relative configuration in the model, the positions of some side chains must be changed. In the predicted model [7] (see Fig. 3) the side chain of Leu 103_H completely screens Tyr 34_L from the binding site. This would have two conse-

quences for the perturbations in the NMR spectrum on binding hapten. There would be a large change in the aliphatic
region, involving downfield shifts on 8 protons from Leu
103_H and the shift on H(3) would have to arise solely from
Trp 93_L. The expected shift for H(3) would then be far
below the observed range of values of 1.3–1.7 ppm. Furthermore haptens without long side chains such as DnpO$^-$ perturb very little in the aliphatic region of the NMR
spectrum on binding so that Leu 103_H cannot define the
lower limit of the combining site. This necessitates chang-
ing the conformation of the H3 loop from residue 102 to
104, so that the leucine side chain is removed from the
site and Tyr 34_L can make contact with the hapten. The
observation of a nuclear Overhauser effect [19] between
H(3) and an aromatic protein resonance at 6.6 ppm provides
another distance to fix the position of H(3) relative to
Tyr 34_L. The proton-proton separation must be less than
3Å to account for the size of the effect.

There are now several further changes necessary as a
consequence of fixing the hapten position relative to
Tyr 34_L. The side chain of Phe 34_H must be rotated about
both Cα–Cβ and Cβ–Cγ bonds so that it is further towards
the rear of the combining site, where it then interacts
unfavourably with Asp 99_H. This latter residue, whose
presence in the combining site has been questioned [7] because of its highly polar nature, is then re-oriented so
that the side chain is exposed to the solvent.

Since some of the perturbed aromatic resonances have
chemical shifts to much higher field than those for free
tyrosine, phenylalanine and tryptophan (Fig. 10), it is
necessary for there to be further aromatic rings affecting
Phe 34_H, Trp 93_L and Tyr 34_L. Interaction between these
residues is small, for the relative geometry required for
hapten binding, so that there must be a second layer of
aromatics perturbing the first. As these second layer
residues are not themselves probed by NMR, tentative

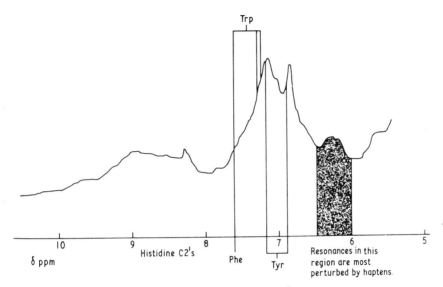

Fig. 10. The aromatic region of the 270 MHz ^{1}H NMR spectrum of Fv, showing the normal resonance positions of aromatic amino acids and the strongly upfield shifted resonances perturbed by hapten binding.

assignments can only be made from the predicted model, which shows that Tyr 104_H can now interact with Tyr 34_L, Phe 50_H with Phe 34_H, and Phe 98_L with Trp 93_L. It is of particular importance, however, that the last pair of interactions will only result in large upfield shifts on the indole resonances if the ring is rotated for the position in the predicted combining site by $180°$ about the Cβ—Cγ bond so that it points into the combining site. The resulting modified combining site with Dnp-glycine bound in it is shown in Fig. 11. The orientation of Dnp glycine in the combining site enables hydrogen bonds to be formed between the nitro groups of the hapten and the NH_2 of Asn 36_L and the OH of Tyr 34_L. For other Dnp haptens the values of the shifts on H(3), H(5) and H(6) differ slightly, but can be accounted for by very small changes in position of the hapten [15].

   Although the correctness of the geometry given here

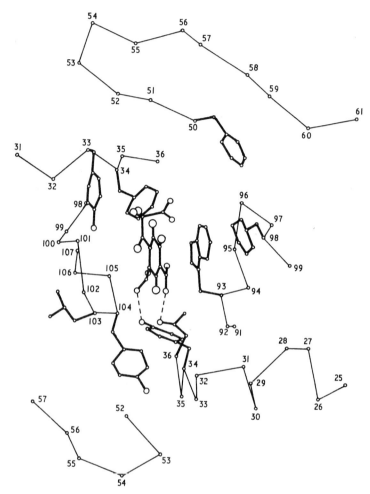

Fig. 11. The combining site of protein 315 refined from the ESR and NMR data, with a Dnp glycine hapten bound in a position parallel to Trp 93_L and capable of hydrogen bonding to Tyr 34_L and Asn 36_L.

depends on the validity of using the Johnson-Bovey treatment, any underestimate by this theory would only lead to
very small movements ($\ll 0.05$ nm) of the rings relative to
the Dnp ring. It should be stressed that, using Haigh and
Mallion's tables, it is impossible to account for the
magnitude of the observed shifts in terms of a physically

possible geometry.

Dependence of hapten orientation on binding affinity

The Dnp haptens that have been studied by NMR so far
have a range of binding constants which extends from 1 mM,
for DnpO$^-$, to 12 μm, for Dnp glycine. All of these haptens
fulfil the conditions for fast exchange, and the haptens
can therefore be easily followed by suitable titrations
and the bound shifts obtained. As the binding becomes
tighter this is no longer true and the determination of
the bound hapten shifts is more difficult. In these circumstances, cross saturation experiments provide a method
of determining directly the bound shifts. Our initial
experiments have been carried out on Nα-acetyl-Nε-(Tnp)-
lysine (K_D ~ 0.2 μM). The method involves irradiating the
bound resonance positions and observing a decrease in
intensity of the free resonance. This decrease arises
from the transfer of saturation from the irradiated bound
resonances to the free resonance. The critical condition
is that the exchange rate is the same as, or greater than
the relaxation rate. By irradiating throughout the aromatic region the bound positions can be found, since they
will correspond to those positions at which maximum peak
height reduction is observed. Two bound positions were
identified, corresponding to upfield shifts of 1.4 and
2.2 ppm (Fig. 12). Although H(3) and H(5) cannot be distinguished, the shifts are very similar to those observed
for the H(3) and H(5) protons of Dnp glycine, Dnp asp. and
DnpO$^-$. Only very minor changes are therefore necessary to
accommodate haptens, of widely differing affinities, with
the orientation of hapten as shown. There is thus a
great similarity in the orientation of the aromatic ring
for all the haptens so far studied, despite different side
chains and changes in binding affinity over a range of
5×10^3. The differing affinities observed therefore
directly reflect the contribution from the hapten side

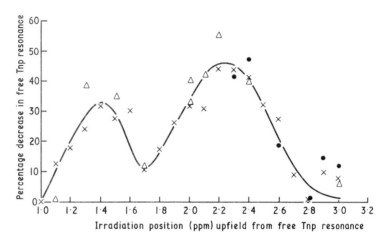

Fig. 12. Reduction in height of the aromatic resonance of free Nα-acetyl-Nε-(Tnp)-lysine upon irradiation at positions upfield from this resonance. The two maxima correspond to resonance positions of the two Tnp aromatic protons bound to Fv. Spectra were recorded at 270 MHz, pH 6.9 and 303 K.

chain to the total binding energy of the hapten to Fv.

Further support for this mode of binding comes from affinity labelling studies, using reactive Dnp haptens. By varying the side chain length, Givol and co-workers were able to preferentially label Tyr 34_L or Lys 52_H using bromoacetyl groups [20]. However in all cases both residues were labelled, but the percentage to each was sensitive to the side chain length. With m-diazonium nitrobenzene, however, Goetzl and Metzger [21] showed that there was exclusive labelling of Tyr 34_L. Such discrimination suggests that there is little rearrangement after the non-covalent binding step before reaction occurs, otherwise other labile groups might be attacked. The labelling of Tyr 34_L rather than Tyr 33_H strongly supports Tyr 34_L as the residue in contact with the 2 nitro group, rather than Phe 34_H. The conclusions on the specificity of protein 315 for dinitrophenyl ligands are that the Dnp ring forms a stacking complex with tryptophan-93_L, and this interaction, which occurs in an aromatic box formed of tryptophan-93$_L$, phenylalanine-34$_H$ and tyrosine-34$_L$, results in highly favourable contributions to the free energy of binding. There is also the possibility of enthalpy contributions from hydrogen bonds involving the two nitro groups and asparagine-36$_L$ and tyrosine-34$_L$. The binding energy contributions from the Dnp side chains may reflect their transfer into the non-polar site. This is illustrated by comparing the affinity of Dnp-O$^-$ (approx. 1 mM) with Dnp-aspartate or Dnp-glycine (approx. 10 μM). The NMR studies show that in each case the Dnp ring binds in a very similar manner. The difference in affinity mainly reflects the substitution of neutral NH in Dnp-aspartate for the charged O$^-$ of Dnp-O$^-$. Additionally, in haptens carrying charged tail groups, there is the possibility of electrostatic interactions with the charged groups at the entrance of the site (aspartate-101$_H$, lysine-52$_H$, arginine-95$_L$ and histidine-102$_H$).

Cross Reaction with Non-polynitrophenyl Ligands

In addition to binding Dnp and Tnp haptens, protein 315 binds menadione. It has been suggested that there are quite distinct binding sites for these different haptens. Work in this laboratory has now shown that it binds many other aromatic molecules, with affinities as high as the original Dnp haptens (Table 5). For menadione, dinitronaphthol-sulphonic acid, ethidium bromide and proflavin, preliminary hapten titrations have been performed to determine the NMR chemical shift changes caused by the protein on the hapten. In general the shifts are upfield and are of comparable magnitude to those observed for Dnp haptens. Since the same protein resonances are also perturbed as with nitrophenyl haptens, it suggests that the same binding site is involved with the same contact residues.

This multispecificity is, however, apparent rather than real. In the cases studied it is the structural similarities of the haptens outweighing other factors which results

TABLE 5

Dissociation Constants of non–Dnp and Tnp haptens
a) Menadione and Dnp-related haptens with strong binding
b) Very weakly binding haptens
c) Novel haptens with medium-strong affinities

in it still binding, rather than some novel property in
each case determining a new mode of interaction with the
protein side chains. Thus, apart from tetracyanoethylene,
all are aromatic molecules. These will favour a hydrophobic
environment. However all are polar or polarisable molecules, capable of dipole-dipole interactions and have potential hydrogen bonding groups i.e. CN, NO_2, CO, OH, NH_2.
Thus all the properties of the Dnps are duplicated in whole
or in part in these molecules. The main difference lies in
the greater size, especially of ethidium bromide, compared
to the Dnp ring. However, it should be realised that the
great variation in side chain length for different Dnp haptens that have been shown to bind, means that the longest,
Dnp lysine, is of a comparable length to ethidium bromide.

Antibody Specificity and Diversity

If individual antibodies are indeed monospecific and
their binding of slightly different haptens is a manifes-
tation of the same specificity, then there should be a
structural basis for this, determined by the number and
orientation of amino acid side chains in the hypervariable
regions forming the combining site. In this way the specificity is not determined solely by the presence or absence
of a single amino acid but rather by several contact residues.

For both the Fv and Fab fragments of protein 315, the
light and heavy chains have been separated and shown to
have different properties. The light chains, either as V_L
or light chain dimers, still have Dnp specificity, in contrast to the heavy chain. Even though the binding is generally weaker than with Fv, there is still a particular
amino acid distribution present on the light chain which
must be necessary for the recognition of Dnp groups,
whether in Fv or in light chain dimers. This is consistent
with the mode of hapten binding in the Fv in which the main
interactions are attributable to light chain residues.

P. GETTINS and R.A. DWEK

Furthermore, the similar binding constants for a range of
Dnp haptens with different side chains to V_L dimer [22]
for which the affinities to Fv vary by a factor of 10^3
(Table 6) suggests that the role of the heavy chain involves mainly side chain interactions. However, such interactions do not determine the specificity of Fv for the Dnp
ring, but rather cause some change in total binding energy.

TABLE 6

Binding of various ligands to protein 315
IgA and its L and V_L fragments

Ligand	$K_a(M^{-1} \times 10^{-3})$		IgA
	L	V_L	
Dnp OH	0.38	0.46	3
Dnp NHCH$_2$COOH	1.1	1.60	50
Dnp NHCH$_2$CH$_2$CH$_2$COOH$_2$	1.2	1.40	5200
Dnp NHCH$_2$CH$_2$CH$_2$NH	1.0	1.20	870
Dnp NHCH$_2$CH$_2$CH$_2$CH(NH$_2$)COOH	1.6	2.0	3400
Dnp NHCH$_2$CH$_2$CH$_2$CH$_2$CH$_2$COOH	2.4	2.2	3600
Dnp NHCH$_2$CH$_2$CH$_2$CH$_2$CH(NH$_2$)COOH	2.5	2.4	1700
Dinitronaphthol 7-sulphonate	3.8	4.5	10
Menadione	2.3	1.5	70

The crystal structures of proteins New and 603 with
bound hapten, have been determined at high resolution, so
that the contact residues are well defined. A comparison
of their quite different specificities with that of protein
315 should be very useful because of the large sequence
homology of all three, which even extends to regions of the
combining site. Protein New binds both vitamin K_1OH and
menadione. The latter hapten lacking the phytyl chain of
the former, results in much weaker binding of menadione
mainly from loss of side chain interactions with the H

chain combining sites. Table 7 compares the hypervariable
residues of proteins 315, 603 and New, with hapten contact
residues in bold type. This shows that Trp 93_L, which has
been implicated as being very important in the specificity
of protein 315 for Dnp, has been replaced by tyrosine in
protein New. Since a stacking interaction is involved in
both combining sites, between Dnp and tryptophan in protein 315 and between the naphthoquinone ring and tyrosine
in protein New, it might be expected that protein New
would bind Dnp haptens, since protein 315 also binds menadione. No Dnp binding has yet been found which suggests
that, while the change from Trp 93_L to Tyr 90_L may be
important in the change in specificity, there are other
factors, such as the presence of hydrogen bonding amino
acid side chains which help distinguish between menadione
and Dnp.

The completely different specificity of protein 603,
for phosphorylcholine, requires a greatly modified combining site despite the high degree of homology between
the framework residues. Replacement of Asn 36_H by Glu 35_H
and of Tyr 58_H by Glu 59_H produces a negatively charged
site capable of interacting with the positively charged
trimethylammonium moiety, while the replacement of Asp 54_H
by Lys 54_H places two positively charged groups close to
the negative phosphoryl group. Unfortunately the full
light chain sequence is not known so that the residue corresponding to Trp 93_L is at present unknown. However, the
removal in protein 603 of the other residues necessary for
Dnp specificity (Asn 36_L changed to alanine and Tyr 34_L to
phenylalanine) would suffice to explain the lack of observed binding of Dnp to it.

Thus the idea that a common three dimensional structure
for the fragments of antibodies can still accommodate completely different specificities by antigen-combining site
complementarity is seen to be valid for the cases considered. The techniques of magnetic resonance coupled with

P. GETTINS and R.A. DWEK

TABLE 7

*Comparison of Combining Site Residues which Interact with Ligands**

Protein	Ligand	L1 (32)	L3 (92)
MOPC315	Dmp	Ser Asn TYR Ala ASN	Leu TRP Phe Arg Asx His Asp
McPC603	Phosphorylcholine	Lys Asx Phe Leu Ala	X X X X X X X
New	Vitamin K	GLY ASN His Val Ser	Ser TYR Asp Arg SER LEU ARG

Protein	H1 (31)	H2 (50 ... 59)	H3 (99)
MOPC315	Ser Gly TYR PHE Trp Asn	Phe Ile LYS Tyr Asp	Asp Asn Asp His Leu Tyr
McPC603	Asp Phe TYR Met GLU	Arg Ser ARG Asn LYS.....GLU	ASN Tyr Tyr Gly Ser Thr TRP Tyr
New	Asn Val Tyr Ile Val	TYR Val Phe Tyr His	Asx LEU ILE ALA Gly Cys

*Interacting residues (capitals) are according to Poljak *et al.* and Rudikoff *et al.* for New and M603 respectively. H1, H2, H3, L1, L2 and L3 are hypervariable regions of H and L chains. X represents an unknown residue due to incomplete sequence.

model building described in this paper should thus enable other antibody systems with a range of specificities to be investigated.

Acknowledgements

We thank our colleagues S.K. Dower, R. Jackson, B.J. Sutton, S. Wain-Hobson and K.J. Willan for their contributions to this work, which forms part of a long term project in this laboratory to investigate antibodies by NMR in collaboration with Professor D. Givol at the Weizmann Institute, Israel. We gratefully acknowledge the continuous help, advice and suggestions of Professor R.R. Porter and Professor R.J.P. Williams. We are also grateful to Dr. Iain Campbell for help with the NMR studies.

We thank the Medical Research Council and Science Research Council for financial support. R.A.D. is a member of the Oxford Enzyme Group.

References

1. Fleischman, J.B., Pain, R.H., and Porter, R.R. (1962). *Arch. Biochem. Biophys. Suppl.* **1**, 174.
2. Edelman, G.M., Cunningham, B.A., Gall, W.E., Gottlieb, P.D., Rutishauser, U. and Waxdal, M.J. (1969). *Proc. Nat. Acad. Sci. U.S.* **63**, 78.
3. Poljak, R.J., Amzel, L.M., Avey, H.P., Becka, L.N. and Nisonoff, A. (1972). *Nature New Biol.* **235**, 137.
4. Rudikoff, S., Potter, M., Segal, P.M., Padlan, E.A. and Davies, D.R. (1972). *Proc. Nat. Acad. Sci. U.S.* **69**, 3689.
5. Schiffer, M., Girling, R.L., Ely, K.R. and Edmundson, A.B. (1973). *Biochem.* **12**, 4620.
6. Epp, O., Colman, P., Fehlhammer, H., Bode, W., Schiffer, M. and Huber, R. (1974). *Eur. J. Biochem.* **45**, 513.
7. Padlan, E.A., Davies, D.R., Pecht, I., Givol, D. and Wright, C.E. (1976). Cold Spring Harbour Symp. on Quant. Biol. in press.
8. Sutton, B.J., Dwek, R.A., Gettins, P., Marsh, D., Wain-Hobson, S., Willan, K. and Givol, D. (1977). *Biochem. J.* in press.
9. Dwek, R.A., Jones, R., Marsh, D., McLaughlin, A.C., Press, E.M., Prich, N.C., and White, A.I. (1975). *Phil. Trans. Royal Soc.* **B 272**, 53.
10. Wain-Hobson, S., Dower, S.K., Gettins, P., Givol, D., McLaughlin, A.C., Pecht, I., Sunderland, C.A. and Dwek, R.A. (1977). *Biochem. J.* in press.
11. Klostergaard, J., Krausz, L.M., Grossberg, A.L., and Pressman, D. (1977). *Immunochemistry* **14**, 107.

12. Eisen, H.N., Simms, E.S. and Porter, M. (1968). *Biochemistry* **7**, 4126.
13. Rockey, J.H., Montgomery, P.C., Underdown, B.J. and Dorrington, K.J. (1972). *Biochem.* **11**, 3172.
14. Freed, R.M., Rockey, J.H. and Davis, R.C. (1976). *Immunochem.* **13**, 509.
15. Dower, S.K., Wain-Hobson, S., Gettins, P., Givol, D., Jackson, R., Perkins, S.J., Sunderland, C., Sutton, B., Wright, C. and Dwek, R.A. (1977). *Biochem. J.* in press.
16. Johnson, C.E. and Bovey, F.A. (1958). *J. Chem. Phys.* **24**, 1012.
17. Haigh, C.W. and Mallion, R.B. (1971). *Molecular Physics* **22**, 955.
18. Hanson, A.W. (1964). *Acta Cryst.* **17**, 559.
19. Noggle, J.H. and Schirmer, R.E. (1971). *The Nuclear Overhauser Effect.* Academic Press, N.Y.
20. For a recent review see Givol, D. (1974). *Essay in Biochemistry* **10**, 73.
21. Goetzl, E.J. and Metzger, H. (1970). *Biochem.* **9**, 3862.
22. Gavish, M., Dwek, R.A. and Givol, D. (1977). *Biochem.* in press.

FLEXIBILITY IN THE STRUCTURE OF TROPOMYOSIN

BRIAN F.P. EDWARDS AND BRIAN D. SYKES

Department of Biochemistry and
MRC Group on Protein Structure and Function
University of Alberta, Edmonton, Alberta, Canada T6G 2H7

Introduction

Tropomyosin is a fibrous protein involved in the thin fil-
ament based Ca^{++} regulation of skeletal muscle [1]. In its
monomeric form in high ionic strength solutions, rabbit
skeletal tropomyosin is composed of two polypeptide chains
each 284 residues long, giving a combined molecular weight
of 66,000 daltons. At low ionic strength tropomyosin is
very highly polymerized in a head to tail fashion. While
the exact three dimensional structure of tropomyosin has
not yet been determined by X-ray crystallographic methods,
the recent determination of the amino acid sequence of α
tropomyosin [2] and the over 90 per cent helical content
seen by CD/ORD provide strong support for a coiled coil
structure of the type originally predicted by Crick [3].
In the coiled coil, each polypeptide chain forms an α
helix with a 3.6 residue repeat, and then the two helices
running in the same direction are coiled around each other.
This structure places every consecutive group of seven
residues in structurally equivalent positions. The sequence shows a regular pattern of hydrophobic residues repeating alternatively at intervals of three and four residues which are proposed to interlock and stabilize the
interface between the two helices of the coiled coil.
Looking at the seven residue repeat pattern in more detail

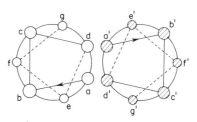

from the end of the coiled coil [4] we see that the interlocking hydrophobic residues occupying positions a,d' and
a'd in the two chains are flanked by the e',g and e,g'
positions which appear statistically as salt linked charged residues. Positions b,c, and f are exposed residues.
With the final evidence obtained from interchain sulphy-
dryl crosslinking that the two chains are in register
[5,6] and not staggered by some multiple of 7 residues,
we have a rod-like molecule of approximate dimensions
20Å × 400Å with a two fold axes of symmetry parallel to
the long axis [7,8].

We endeavour in this manuscript to provide additional
information on the nature of the structure of tropomyosin
using high resolution nuclear magnetic resonance methods.
Recently, NMR linewidth and relaxation time measurements
have been used to determine the magnitude of internal
motions in proteins. Of specific interest in this manuscript is the range of internal motions possible in the
coiled coil structure. From an NMR experiment only the
linewidth measurement addresses this question; the spin
relaxation time measurement is dominated by spin diffusion
for large molecules and high NMR frequencies [9,10,11].
Normally the measured linewidth in the presence of internal motion is a function of four parameters: the overall
correlation time, τ_c, which can be estimated from the
Stokes–Einstein equation or hydrodynamic experiments; the
internuclear distance, r, which can be estimated from
molecular models; the geometry (angles) of the internal
motion relative to the overall motion, which is often ill-
defined; and the correlation time of the internal motion,

τ_i, which is the unknown we wish to measure. The regular structure of tropomyosin allows the angles of internal motion axis to be determined relative to the moment of intertia tensor of the molecule [12]. Reasonable estimates of the range of internal motions can then be made.

NMR Theory and Considerations

For the purposes of the NMR we can mask the amino acid sequence and look only at the aromatic amino acids. The 'NMR sequence' is shown in Fig. 1. As we can see 5 out of

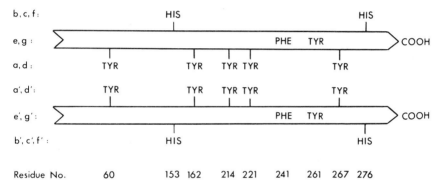

Fig. 1. Diagrammatic representation of the sequence of tropomyosin showing only aromatic amino acids.

the 6 tyrosines occupy the a,d interlocking hydrophobic positions, 1 tyrosine and 1 phenylalanine occupy the flanking e,g positions and the 2 histidines are in the exposed f position. We are particularly interested in the 5 tyrosines in the interface of the coiled coil since the rest of the residues will be quite exposed and free to undergo a high degree of internal motion.

To calculate linewidths and T_1's, tyrosines on the coiled coil can be represented by tyrosines on a rod (see Fig. 2). The rotational diffusion axis are assumed to be oriented with z parallel to the long axis of the coiled coil and the rotational diffusion constants D_x and D_y are assumed equal. Since the H(meta)—H(ortho) internuclear vector is parallel to the C_β-ring bond axis, rotation

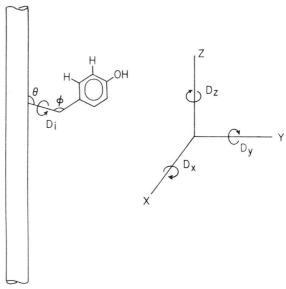

Fig. 2. Representation of tropomyosin as a rod, defining the rotational diffusion axis and angles for internal motion of tyrosine.

about this axis is invisible in the NMR calculation (ineffective in narrowing the NMR resonances). We will therefore consider only internal motion about the C_α–C_β axis with diffusion constant D_i. We will also neglect other (inter-residue) internuclear interactions for the moment.

The formula for $1/T_1$ and $1/T_2$ can then be obtained from the results of Levine *et al.* [13] or London and Airtabile [14]

$$\frac{1}{T_1} = \frac{3}{10} \frac{\gamma^4 \hbar^2}{r^6} \{J(\omega) + 4J(2\omega)\}$$

$$\frac{1}{T_2} = \frac{3}{20} \frac{\gamma^4 \hbar^2}{r^6} \{3J(o) + 5J(\omega) + 2J(2\omega)\}$$

with $\dfrac{3}{20} \dfrac{\gamma^4 \hbar^2}{r^6} = 7.52 \times 10^8$ for ^{1}H with r = 2.5Å

and $J(\omega) = \displaystyle\sum_{z,a} |d^2_{za}(\vartheta)|^2 |d^2_{ao}(\varphi)|^2 \tau^{az}$

where the indices z and a run from -2 to 2, the $d^2m,m'(\beta)$ are reduced Wigner rotation matrices [15], and

$$\tau^{az} = \frac{\tau^*}{1+(\omega\tau^*)^2}$$

where

$$\tau^* = \frac{1}{E_z + a^2 D_i}$$

and

$$E_o = \frac{1}{6D_x}$$

$$E_{\pm 1} = \frac{1}{5D_x + D_z}$$

$$E_{\pm 2} = \frac{1}{2D_x + 4D_z}$$

The rotational diffusion constants for a prolate ellipsoid of rotation with axis a,b,b can be calculated from Perrin's equations [16]

$$D_x = \frac{3kT}{16\pi\eta a^3}\left[1-2\ln\frac{(2a)}{b}\right]$$

$$D_x + D_z = \frac{3kT}{16\pi\eta ab^2}\left[1-\frac{b^2}{a^2}\ln\frac{(2a)}{b}\right]$$

for the case a $\gg$ b. Converting our cylinder of dimensions 400Å in length and 20Å in diameter into a prolate ellipsoid of equivalent volume (a = 200Å, b = 12.2Å), and taking the viscosity of water as 10^{-2} poise one obtains

$$D_x = 1.9 \times 10^5 \ sec^{-1}$$

$$D_z = 8.3 \times 10^6 \ sec^{-1}$$

By comparison, a sphere of equivalent volume has a radius of 31Å, and a rotational diffusion constant D = 5.7×10^6 sec^{-1} which converts to a rotational correlation time τ_c = 1/6D = 29 nsec.

The calculated linewidth at 270 MHz as a function of

Θ and D_i (for $\varphi = 109°28'$) is presented in Fig. 3. The geometry is such that the linewidths increase with increasing angle Θ for $D_i \rightarrow 0$, but the magnitude of the line-narrowing for large D_i is greater for larger Θ's. This reflects the fact that for small angles ϑ the H—H vector will be nearly perpendicular to the long axis of the rod so that the faster D_z overall motion will narrow the line

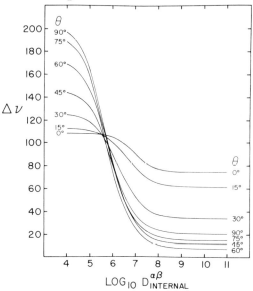

Fig. 3. Calculated linewidths for tyrosine on an idealized rod-like molecule as a function of the rate of internal rotation for various angles of attachment to the rod.

and the internal motion will not add much to this; whereas for large angles Θ, the H—H vector will be nearly parallel to the long axis so that its orientation is not greatly averaged by the rotation about the long axis but internal motion perpendicular to the long axis is very effective in averaging out the dipolar interaction.

We will repeat at this stage that individual T_1's are considered meaningless given the dominance of cross relaxation [9,10,11] (spin diffusion) in spin lattice relaxation of large proteins at high fields. The shortest possible value of T_1 for any angle ϑ and value of D_i is ~1 sec for

a rod of the present dimensions. The longest possible
value of the cross relaxation time, however, will be close
to $4/5$ T_2 ~.2 sec. (T_1 is observed to be equal across the
aromatic spectrum supporting the above conclusions [17].)

The final consideration is the angle ϑ. A standard α
helix was calculated (poly ALA) with (φ,ψ) angles of
$(-57.37°,-47.52°)$. The $C_\alpha-C_\beta$ bond was found to make an
angle of $59°$ with the long axis of the helix. Given that
the two helices in tropomyosin cross each other at $20°$,
the angle ϑ can take values between $49°$ and $69°$ for the
perfectly regular structure, being nearer $69°$ for the
interface tyrosine positions.

Experimental

NMR measurements were made at 270 MHz on a Bruker
HXS–270 NMR spectrometer operating in the Fourier transform mode. Typical acquisition parameters were acquisition time 0.5 sec, sweep width 4000 Hz, 4K points, and
1 Hz linebroadening. The spectra in this paper were all
taken at 27–28°C. All pH's were measured with a glass
electrode standardized in H_2O buffers; they are uncorrected for D_2O. α-Tropomyosin was isolated from rabbit skeletal muscle by the method of Cummins and Perry [18].

Results

The aromatic portion of the 270 MHz NMR spectrum of
tropomyosin at pH 8 and high ionic strength is shown in
Fig. 4. For the purpose of this manuscript we will assign
the spectrum as follows, although only the histidine C2
and C4 protons have been assigned to particular amino acid
residues [17]: $\delta \approx 6.8$ ppm downfield from DSS, ortho
tyrosine protons (12); $\delta \approx 6.9$ ppm, C4 histidine protons
(2); $\delta \approx 7.3$ ppm, meta tyrosine protons (12) and phenylalanine protons (5); $\delta \approx 7.7$ ppm, C2 histidine protons
(2); and $\delta \approx 8.5$ ppm, formate ion. Shown for comparison is
the 270 MHz NMR spectrum of the bovine pancreatic trypsin

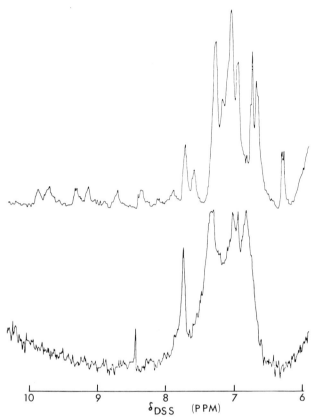

Fig. 4. A comparison between the aromatic region of the 270 MHz ^{1}H NMR spectra of BPTI ($\sim$ 2.5 mM in PIPES buffer, pH 7.72) and of tropomyosin (18 mg/ml in 40 mM KPO$_4$, 1 M KCl, 0.5 mM EDTA, 0.25 mM DSS, D$_2$O, pH = 7.95).

inhibitor (BPTI) at a similar pH. BPTI was chosen for comparison because it has only tyrosine and phenylalanine and because, being small, it resembles the thin coiled coil in having very little interior. The assignment of the BPTI spectrum has been made in detail [19]. It is immediately obvious that the spread in chemical shifts for the aromatic ring protons in BPTI ($\delta \approx$ 6.3 to 7.6 ppm) is roughly twice that for tropomyosin. In addition, the individual tyrosines in BPTI are in unique environments reflected by unique chemical shifts. The tropomyosin spectrum by com-

parison looks like that of standard amino acids broadened
by increase in molecular weight. This immediately suggests
that all of the residues are relatively exposed to solvent [20] and none so restricted in a unique environment
as to produce a different chemical shift. As stated earlier we will concentrate on tyrosine protons in this manuscript and the histidine protons, in particular, will be
treated elsewhere [17].

   Fig. 5 compares tropomyosin at pH near 6 and pH near 8.
At pH 6 the histidine C2 protons have moved to $\delta \approx 8.6$ ppm
and C4 protons to $\delta \approx 7.4$ ppm leaving the ortho tyrosine
protons not overlapped by other residues. All of the resonance intensity of the aromatic protons can be accounted

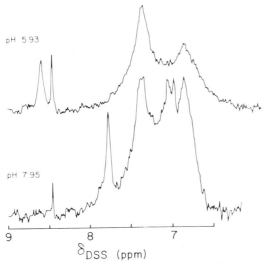

Fig. 5. The aromatic resonances of the 270 MHz ^{1}H NMR spectrum of α-
tropomyosin as a function of pH. Both spectra were taken on the same
tropomyosin samples used for Fig. 4.

for when integrated with respect to the histidine C2 protons. We can estimate a linewidth for the ortho tyrosine
protons. (The meta tyrosine protons interact not only with
the ortho tyrosine protons but also with the β,β' protons
and therefore are not discussed.) At pH near 8 the linewidth is approximately 60 Hz. When compared with the

calculated linewidth of 150-170 Hz for ϑ in the range
50°—70° and $D_i \to 0$, this implies considerable internal motion with resultant line narrowing. At pH near 6, the linewidth increases to approximately 112 Hz. While this could
reflect a tightening of the structure with concomitant
decrease in internal flexibility, other evidence [17] suggested that the increase in linewidth could also be due to
an increase in polymerization favoured by the protonation
of histidine 276 which is involved in a salt link with
aspartate 2 in the overlap region [4].

The coiled coil can be stabilised by going to low pH
[21]. Tropomyosin is also monomeric under these conditions. The spectrum of tropomyosin at pH 3.9 is shown in
Fig. 6. Here the spectrum has the appearance of the histidine C2 protons riding on top of the now slowly exchanging
NH protons, and of relatively narrow single phenylalanine
and tyrosine resonances appearing on top of underlying

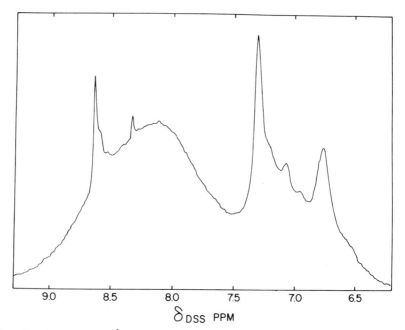

Fig. 6. The 270 MHz ^{1}H NMR spectrum of the aromatic residues of tropomyosin at low pH (20 mg/ml in 13 mM DC1, pH 3.9).

broader aromatic resonances. The narrower phenylalanine
and tyrosine resonances correspond to the phenylalanine
241 and tyrosine 261 in the more exposed flank positions
whereas the broader resonances correspond to the now
immobilised interface tyrosines with calculated linewidths
near 200 Hz.

The spectrum of tropomyosin highly polymerised in low
ionic strength solution at pH 8 is shown in Fig. 7. Here

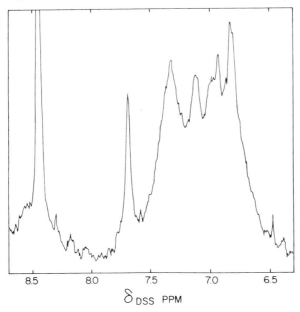

Fig. 7. The 270 MHz ^{1}H NMR spectra of the aromatic residues of tropomyosin at low ionic strength (20 mg/ml in 10 mM KPO , 0.5 mM EDTA,
0.25 M DSS. pH 8.80).

the viscosity of the solution is so high that the tube can
easily be inverted without the solution running down the
walls. The mere observation of the spectrum is remarkable.
Yet, all of the resonance intensity of the aromatic portion is present when integrated relative to the histidine
C2 protons. Since the calculated linewidths would be much
larger for the polymeric tropomyosin, the observed narrow
lines demand flexibility and internal motion for all of
the sidechains of tropomyosin even in the interface

position.

Conclusions

The coiled coil structure of tropomyosin near neutral
pH is a flexible open structure in both the monomeric and
polymerised forms. The presumed regularity of the structure is not reflected in rigidity in the structure. While
the linewidth calculations in this paper have been para-
meterised in terms of narrowing caused by motion around
the $C_\alpha - C_\beta$ bond, it is of course possible that other modes
of motion are possible. In particular, fixing the (φ, ψ)
angles for tyrosine at those for an α-helix restricts the
motion of the γ carbon caused by rotation about the α, β
bond. The internal motion about the α, β bond axis is
therefore more realistically a restricted or limited
oscillation. An indistinguishable alternative is flexibility in the long axis of tropomyosin. Assuming a different
mode of motion, however, changes only the quantitative
aspects of the calculation, and not the central conclusion that large scale motions are present in tropomyosin.

Acknowledgements

We have benefited greatly from consultations with Dr.
L. Smillie and the members of his laboratory. They provided the tropomyosin and guided us in the various biochemical techniques which we applied to it. We are also
indebted to Dr. Anita Sielecki for the bond angle calculations on poly ALA. This research was supported by the
Medical Research Council of Canada.

References

1. Cohen, C. (1975). *Sci. Amer.* **233**, 36.
2. Stone, D., Sodek, J., Johnson, P. and Smillie, L.B. (1974). *Proc. IX FEBS Meeting Budapest* **31**, 125.
3. Crick, F.H.C. (1953). *Acta Cryst.* **6**, 689.
4. McLachlan, A.D. and Stewart, M. (1976). *J. Mol. Biol.* **98**, 293.
5. Stewart, M. (1975). *FEBS Letters* **53**, 5.
6. Johnson, P. and Smillie, L.B. (1975). *Biochem. Biophys. Res.*

Commun. **64**, 1316.

7. Casper, D.L.D., Cohen, C. and Longley, W. (1969). *J. Mol. Biol.*
    41, 87.
8. Longley, W. (1977). *Intl. J. Peptide Protein Res.* **9**, 49.
9. Hull, W.E. and Sykes, B.D. (1975). *J. Chem. Phys.* **63**, 867.
10. Kalk, A. and Berendsen, H.J.C. (1976). *J. Mag. Res.* **24**, 343.
11. Sykes, B.D., Hull, W.E. and Snyder, G.H. (1977). *Biophysical J.*
    (in press).
12. Huntress, W.T. (1968). *J. Chem. Phys.* **48**, 3524.
13. Levine, Y.K., Birdsall, N.J.M., Lee, A.G., Metcalfe, J.C.,
    Partington, P. and Roberts, C.G.K. (1974). *J. Chem. Phys.* **60**,
    2890.
14. London, R.E. and Airtabile, J. (1976). *J. Chem. Phys.* **65**, 2443.
15. Brink, D.M. and Satchler, G.R. (1975). *Angular Momentum.* Claren-
    don Press, Oxford.
16. Edsall, J.T. (1953). *In The Proteins*, ed. H. Neurath and K.
    Bailey, Vol.I, Part B 549, Academic Press.
17. Edwards, B.F.P. and Sykes, B.D. (manuscript in preparation).
18. Cummins, P. and Perry, S.V. (1973). *Biochem. J.* **133**, 765.
19. Snyder, G.H., Rowan, R., Karplus, S. and Sykes, B.D. (1975).
    Biochem. **14**, 3765.
20. Lowey, S. (1965). *J. Biol. Chem.* **240**, 2421.
21. Holtzer, A.M. and Naelken, M.E. (1964). *In: Biochemistry of
    Muscle Contraction*, ed. J. Gergeley Little, Brown & Co., Boston.

MULTINUCLEAR NMR APPLIED TO THE STUDY OF THE SOLUTION STRUCTURE OF PROTEIN — NUCLEOTIDE COMPLEXES

JOSEPH E. COLEMAN and IAN M. ARMITAGE

The Department of Molecular Biophysics and Biochemistry,
Yale University and the Section of Physical Sciences,
Yale University School of Medicine

Introduction

Gene 5 of the filamentous bacteriophages fd, f1 or M13 codes for a small single stranded DNA-binding protein of 87 amino acids required for the shift from double-stranded to single-stranded DNA synthesis during the replication of this phage in E. coli [1–3]. After approximately 200 double-stranded replicative forms of the phage have formed, gene 5 is elaborated and functions by forming a stoichiometric complex with newly synthesized viral DNA and prevents its use as a template for synthesis of the complementary strand.

The gene 5-DNA complex then migrates to the cytoplasmic surface of the cell membrane where the gene 5 monomers are replaced by gene 8 monomers (the coat protein contained in the membrane) as the DNA migrates through the membrane [4,5].

A variety of techniques show that each gene 5 monomer complexes about 4 nucleotides in binding cooperatively to cover completely the ~5700 bases of the circular single-stranded viral genome [6,7]. The molecular biology of fd replication and packaging are covered in detail elsewhere [4,5].

The small size of this protein compared to many other DNA binding proteins and its known amino acid sequence

[8,9] make this protein ideal for direct high resolution
NMR studies of protein-nucleotide interactions. Its elaboration in relatively large quantities by a bacterial system also makes it particularly accessible to the methods
required for the isotopic enrichment of specific nuclei of
the protein. Thus the NMR methods applicable to gene 5
protein-nucleotide interactions provide a prototype system
to demonstrate the application of methods in high resolution ^{1}H, ^{19}F, and ^{31}P FT-NMR that can be applied for the
determination of the solution structure of protein-nucleotide complexes. Results already obtained and future approaches are outlined in this paper.

Materials and Methods

Gene 5 Protein

Homogeneous gene 5 protein was prepared by DNA- cellu-
lose chromatography as previously described [10]. For
calculations of the concentration of gene 5 protein, an
$E_{276}^{0.1\%}$ = 0.73 was employed [11]. pH will refer throughout to the direct reading on a radiometer pH meter, model
26, fitted with a GK2023C electrode, taken on either H_2O
or D_2O solutions.

Nucleotides

Tetra- and octanucleotides of defined sequences were
purchased from Collaborative Research (Waltham, Mass.).
Nucleotides were dissolved in 99.8 per cent D_2O and extracted with a CCl_4 solution of dithizone to remove paramagnetic metals. Nucleotides were added to the protein solutions by lyophilizing the correct volume and concentration
of nucleotide from D_2O buffer and adding the lyophilized
nucleotide to the protein solution.

NMR Methods

^{1}H-NMR spectra were recorded on a FT-270 MHz Bruker
spectrometer with an Oxford Instrument Company superconducting magnet coupled to a Nicolet 1085 computer with a 293
pulse controller and a 294 disc system and with a Bruker

HX270 spectrometer interfaced to a Nicolet BNC-12 computer
and a Nicolet 2931/0 controller. D_2O present in the sample
served as a field-frequency lock. Irradiation of the proton resonance of water was carried out by the gated decoupling technique [12]. Measurements were made at 25°C and
chemical shift values are reported as parts per million
(ppm) downfield from sodium 2,2-dimethyl-2-silapentane-5-
sulfonate. Samples were contained in 0.01 M $DPO_4^=$, 98 per
cent D_2O, pH 8.0, 25°C.

^{31}P-NMR spectra were recorded on a FT-Bruker HFX-90
spectrometer operating at 36.4 MHz as previously described
[13]. Measurements were made at 25°C and chemical shifts in
ppm are reported relative to 85 per cent H_3PO_4. Samples
were contained in either 0.01 M Tris-HCl or 0.01 M $HPO_4^=$,
pH 8.0, 25°C.

^{19}F-NMR spectra were recorded on a FT-Bruker HFX-90
spectrometer operating at 84.6 MHz. D_2O present in the
sample served as the field-frequency lock and chemical
shifts are reported in ppm relative to CF_2COOH. Samples
were contained in 0.01 M Tris-HCl, 90 per cent H_2O—10 per
cent D_2O, pH 8.0, 25°C.

Results and Discussion

Basic Information on Gene 5-DNA Interaction

On binding to single-stranded DNA, gene 5 protein
causes striking changes in the base chromophores. The 260
nm absorption bands become hyperchromic, even more so than
in melted doubled-stranded DNA. Accompanying the hyper-
chromia, the ellipticity of the base chromophores undergoes remarkable changes; for fd DNA the $[\vartheta]_{265}$ shifts from
$+1 \times 10^4$ to -1×10^4 deg cm^2/d mole [10]. Using these optical changes to titrate the number of gene 5 monomers
interacting with a specific length of polynucleotide, it
has been determined that each gene 5 monomer covers ~4
bases. One gene 5 monomer binds to a tetranucleotide, e.g.
$d(pA)_4$, with a dissociation of ~10^{-7} M. Two gene 5 mono-

mers bind to an octanucleotide, e.g. $d(pA)_8$, with a dissociation constant $\sim 10^{-8}$ M or less [10,14]. Using the circular dichroism changes induced in the DNA by gene 5 protein as a binding assay it has been found that modification of several functional groups on the protein block
binding to the DNA. Nitration of tyrosyl residues number
26, 41 and 56 block DNA binding. A tetranucleotide completely protects these tyrosyls from nitration [10]. Likewise the single sulfhydryl of Cys 33 is protected from
reaction with Hg(II) by DNA binding. The cysteinyl residue
has also been observed to crosslink to the DNA upon UV
irradiation of the complex [15]. Lysyl residues occur in
the gene 5 polypeptide chain at positions 3, 7, 24, 46, 69
and 87. Acetylation of these six residues prevents DNA
binding. In contrast to the above modifications nucleotide
binding does not prevent acylation of the lysyls which if
carried out on the complex results in complete dissociation [10]. No evidence has been found that the single histidyl, His 64, is involved in nucleotide binding. Alkyla-
tion of the protein with iodoacetate does not alter binding [10]. Arginyl residues have not been chemically modified as yet.

Based on the above results of chemical modification,
several hypotheses can be put forward as to the type of
molecular interaction between protein groups and the nucleotide chain responsible for the tight binding of nucleotides to gene 5 protein. The positively charged lysyl εNH_3^+
groups can be postulated to participate in binding by
neutralizing the negatively charged phosphodiester backbone of the nucleotide, a hypothesis put forward for almost all DNA binding proteins. Whether they do this by
specific hydrogen bond formation with phosphate oxygens
may need to be re-examined in light of the ^{1}H-NMR results
(see below). Tyrosyl residues could participate either in
hydrogen bonding between the phenolic hydroxyl groups and
the bases or by intercalation of the tyrosyl phenyl rings

with bases. A specific chemical interaction between the
single sulfhydryl group and a nucleotide cannot be so
easily visualized. The above postulated specific interactions between nucleotide and protein as well as other
specific and general aspects of the alterations of protein
structure that occur on nucleotide binding can now be
examined by high resolution FT-NMR methods.

^{19}F-NMR of Gene 5 Protein Labelled with m-Fluorotyrosine

^{19}F-NMR studies on the structure and function of a gene
5 protein biosynthetically labelled with *m*-fluorotyrosyl
residues will be discussed first, since this method can
answer in straightforward fashion questions about the environment of these residues in the protein and determine
which ones interact with the nucleotide as well as permit
speculation on the specific chemical nature of the interaction. By using a tyrosine auxotroph of an Hfr strain of
E. coli we have discovered that growth on *m*-fluorotyrosine
does not prevent infection by fd, elaboration of gene 5
protein and the production of new phage. Growth is slowed
and yields are not as great as normal, but this probably
reflects the slow growth resulting from the susceptibility
of *m*-fluorotyrosyl tRNA to hydrolysis. Since gene 5 is an
essential component of the phage replication cycle, the
m-fluorotyrosyl gene 5 protein was concluded to be functional as proved by the later demonstration of binding of
the pure fluorotyrosyl protein to deoxynucleotides. Incorporation of *m*-fluorotyrosine into gene 5 can also be
achieved by suppressing the aromatic amino acid synthetic
pathway in *E. coli* by large amounts of tryptophan and p-
phenylalanine [16].

The ^{19}F-NMR spectrum of homogenous *m*-fluorotyrosyl gene
5 protein is shown in Fig. 1A. There are two downfield
resonances, 2.8 and 2.2 ppm downfield from the chemical
shift expected for free *m*-fluorotyrosine, while three of
the resonances are grouped between 58 and 59 ppm, approxi-

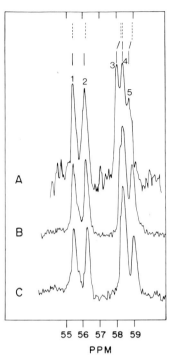

Fig. 1. ^{19}F NMR of *m*-fluorotyrosyl gene 5 protein 7.5 × 10⁻⁴ M (A);
plus 1 equiv. of tetranucleotides of random sequences (B); plus 3
equiv. of tetranucleotides (C); the vertical lines represent the chemical shift positions of the resonances in the free protein (——) and
in the tetranucleotide complex (---) (from reference 14).

mately the chemical shift expected for free *m*-fluorotyro-
sine [14,17,18]. This immediately confirms the conclusion
that two tyrosyl residues of the protein are 'buried'
(Tyr 34 and 61), their downfield shifts reflecting documented effects on the ^{19}F resonances of fluorotyrosyl
residues enclosed within protein structure, with alteration of surrounding dielectric constant, more van der
Waals contacts and relative immobilisation [17–19]. Three
tyrosyl residues must be 'surface' tyrosyls and represent
the three residues subject to nitration (Tyr 26, 41 and
56). It is these three surface tyrosyls which interact
with deoxynucleotides. Their resonances shift upfield when
either tetra or octanucleotides bind (Figures 1B and 2B).

The shift is 0.1 to 0.3 ppm when tetranucleotides of random sequence are used (Fig. 1). With the binding of d(pA)$_8$ one of the surface tyrosyl ^{19}F resonances shifts upfield ~0.8 ppm (Fig. 2, Table 1). A similar pattern of shifts is

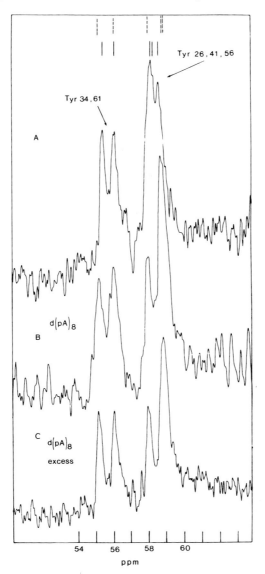

Fig. 2. ^{19}F NMR of *m*-fluorotyrosyl gene 5 protein, .5 × 10^{-3} M (A); plus 1 equiv. of d(pA)$_8$ (B); plus 2 equiv. of d(pA) (C). Vertical lines as in Fig. 1.

TABLE 1

¹⁹F Chemical Shifts in Fluorotyrosyl Gene 5 Protein and its Complexes with d(pA)₈ and d(pT)₄

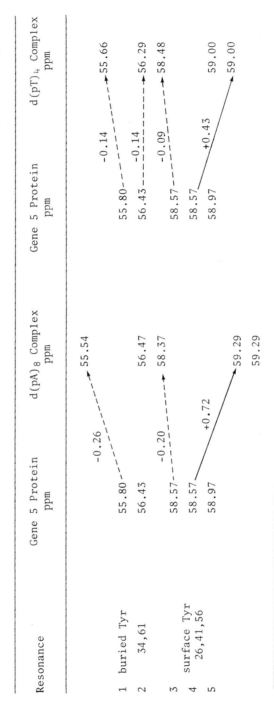

Resonance	Gene 5 Protein ppm	d(pA)₈ Complex ppm	Gene 5 Protein ppm	d(pT)₄ Complex ppm
1 buried Tyr	55.80	55.54	55.80	55.66
2 34,61	56.43	56.47	56.43	56.29
3	58.57	58.37	58.57	58.48
4 surface Tyr 26,41,56	58.57	59.29	58.57	59.00
5	58.97	59.29	58.97	59.00

*Chemical shifts are accurate to ±0.06 ppm. The solid arrows indicate the maximum upfield shifts shown by the fluorines on the surface tyrosyls, while the dotted arrows indicate small downfield shifts that occur on formation of the nucleotide complexes. Conditions: 0.01 M Tris, pH 8.0, 10 per cent D₂O, 25°C.

observed on the binding of d(pT)$_4$, but the maximum upfield shift is now ~0.4 ppm (Fig. 3, Table 1). These upfield shifts are of the right magnitude to represent ring current shifts induced by intercalation of the base rings, and the shifts are greater for adenine than for thymine as expected on theoretical grounds.

The current model of gene 5 protein projects a roughly spherical molecule with a radius of ~14Å [14]. By the

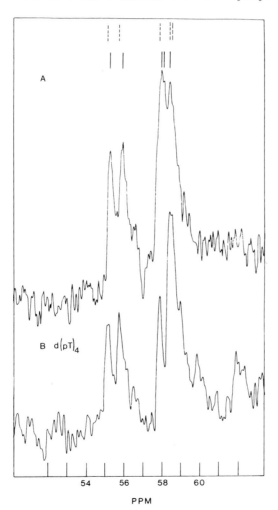

Fig. 3. ^{19}F NMR of *m*-fluorotyrosyl gene 5 protein, .5 × 10^{-3} M (A); plus 2 equiv of d(pT)$_4$. Vertical lines as in Fig. 1.

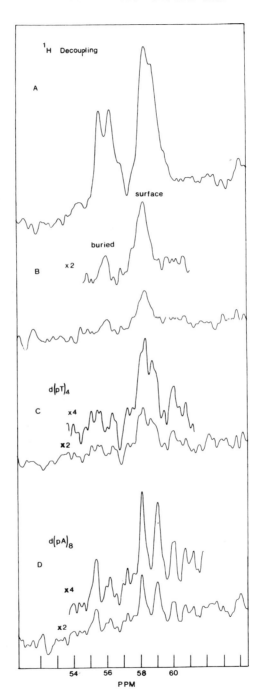

Stokes-Einstein treatment this would have an isotropic tumbling time, τ_c, of ~8.4 × 10^{-9} sec. With this short overall rotational correlation time and assuming all the fluorine relaxation is due to H-F dipolar interactions, immobilised tyrosyls with τ_c close to that of the protein would be expected to show a theoretical Nuclear Overhauser Effect, NOE (η + 1) of 0.2. Thus only 20 per cent or less of the resonance should remain in the presence of proton decoupling. On the other hand, the NOE will be extremely sensitive to internal motion as has been previously reported by Hull and Sykes for fluorotyrosine alkaline phosphatase [18]. For the surface tyrosyls of gene 5, internal rotation about the $C_\alpha - C_\beta$ bond with a τ_i on the order of 10^{-8} to 10^{-10}s will give rise to an observed NOE > 0.2. Thus considerably less resonance attenuation might be expressed for these tyrosyls in the presence of ^{1}H decoupling. Preliminary ^{1}H decoupling experiments do show the above predicted difference in the NOE's of the buried vs. the surface tyrosyls, the latter retaining significantly more resonance in the presence of ^{1}H decoupling (Figs. 4A and 4B).

Binding of either d(pT)$_4$ (Fig. 4C) or d(pA)$_8$ (Fig. 4D) appear to constrain significantly the internal rotation of the surface tyrosyls, since considerably less resonance intensity remains on ^{1}H-decoupling. In the case of the d(pT)$_4$ complex this may be attributed to more immobilisation of the tyrosyl residues by nucleotide binding, since τ_c for the complex would not be expected to increase significantly.

On the other hand, when an octanucleotide binds to gene 5 and knits two monomers together, the long axis of the complex now becomes at least 58Å [14]. Assuming that the

---

Fig. 4. (opposite) ^{19}F NMR of gene 5 protein, .5 × 10^{-3} M (A); the proton decoupled spectrum of the same sample (B); the proton decoupled spectrum of the d(pT)$_4$ complex (C); the proton decoupled spectra of the d(pA)$_8$ complex (D).

slowest rotational correlation time involves the tumbling about an axis perpendicular to this dimension, we can calculate a maximum increase in the applicable correlation time for overall tumbling of the molecule, τ_c, to 6.9×10^{-8} sec. A single correlation time of this magnitude predicts an NOE value such that the [1]H-decoupled resonance should be negative and less than 5 per cent of that in the absence of ^{1}H decoupling. Indeed no resonance is visible at the position of the 'buried' tyrosyls in the $d(pA)_8$ complex (Fig. 4D) compatible with the much larger τ_c. Some slight resonance remains in the [1]H-decoupled spectrum at the position of the 'surface' tyrosyls as it does in the corresponding spectrum for the $d(pT)_4$ complex. The patterns of the remaining resonance in the [1]H-decoupled spectra of the two complexes are different, but the data are too preliminary to tell whether such differences may be used to infer degrees of immobilisation of the several 'surface' tyrosyls. Larger concentrations and sample size are required to answer this and are the subject of continuing investigation. The currently observed expression of the expected [19]F–{H} NOE in the free protein and its nucleotide complexes is compatible with a model in which the nucleotide interacts with and impedes the free rotation of surface tyrosyl residues on the protein.

While the upfield shifts of some of the ^{19}F resonances on the surface fluorotyrosyl residues of gene 5 protein support the single-stranded intercalation model of nucleotide binding to be presented below, other changes in the environment of the fluorotyrosyl residues are likely to accompany the intercalation. Solvent is likely to be excluded from the DNA-binding groove with consequent change in the dielectric constant; van der Waals contacts are also likely to increase. As far as the tyrosyl environment is concerned nucleotide binding may be not unlike folding the surface tyrosyls into the interior of the protein. Such changes have generally been associated with downfield

shifts of the ^{19}F resonances [17–19]. While upfield shifts
appear to predominate on nucleotide binding to fluorotyrosyl gene 5 protein (presumably due to the ring currents
of the intercalated bases), these may not be fully expressed due to environmental changes favouring downfield
shifts. In fact the resonance remaining at the lowfield
position of the 3 surface tyrosyls does shift downfield
0.1 to 0.2 ppm in the $d(pT)_4$ and $d(pA)_8$ complexes respectively (Table 1).

There is a small but significant change, ~0.2 ppm, in
the chemical shift difference between the 'buried' [^{19}F]-
tyrosyls in the $d(pA)_8$ complex (Fig. 2, Table 1), which is
not as pronounced in the tetranucleotide complexes (Figs.
1 and 3). This appears to reflect conformational changes
in the protein induced by the cooperative binding of two
monomers to the octanucleotide, changes also suggested by
the proton spectra to be shown below.

*The 270-MHz ^{1}H NMR of Gene 5 Protein-Octa- and Tetra-
nucleotide Complexes*

The proton spectrum of native gene 5 protein (dialyzed
against 99.8 per cent D_2O–0.01 M $DPO_4^=$) at 270 MHz is
shown in Fig. 5A. The C(2)H resonance of the single histidyl, His 64, is present at 8.02 ppm (peak 1). A number of
overlapping aromatic resonances containing the -CH resonances from the 5 Tyr and 3 Phe residues and the C(4)H of
His 64 appear in the region 6.5 to 7.5 ppm (peaks 3 to 9).
The integrated area under these resonances approximates
the 36 protons expected using the single C(2) histidine
proton as the standard.

In the highfield region of the spectrum the ε-CH$_2$
groups of the 6 Lys residues are resolved in a narrow line
at 3.1 ppm. The δ-CH$_2$ resonances of the 4 Arg residues
occur at 3.2 to 3.5 ppm. A large group of overlapping
resonances from the methyl groups of the aliphatic residues occurs from 0.5 to 2 ppm and these are assigned as

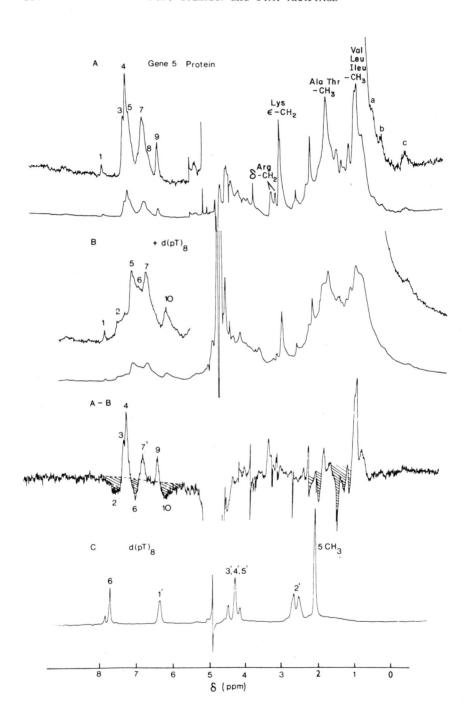

indicated in Fig. 5A. There are a few highfield resonances
near 0 ppm (peaks a, b and c) which represent methyl resonances shifted upfield by ring currents. These assignments are made on the basis of the previous ^{1}H-NMR work on
proteins [20–24].

The 270 MHz ^{1}H spectrum of gene 5 protein in the presence of 1 mole of d(pT)$_8$ per 2 moles of protein is shown
in Fig. 5B. Significant changes occur in both the linewidth and the chemical shift of a number of the resonances
of the aromatic and aliphatic protons of the protein on
formation of the octanucleotide complex. These changes
are best visualised in the difference spectrum A–B. In A–B
the resonances above the baseline are those that have disappeared from the protein on complex formation, while the
inverted crosslined peaks are new resonances appearing in
the complex. The difference spectrum in the aromatic region has been computed by assuming that the resonance of
the C(2)H proton of His 64 is unchanged in the complex
(compare Figs. 5A and 5B). In the aliphatic region of the
spectrum the difference has been computed by assuming that
the magnitude of the resonance from the lysyl εCH_2 groups
is unchanged on complex formation (see below). The spectrum of free d(pT)$_8$ under the same conditions is shown in
Fig. 5C.

The ^{1}H spectrum of the d(pT)$_8$ octanucleotide complex
will serve as a prototype from which to discuss a systema-
tic study of the changes in the resonance of protons on
specific amino acid residues accompanying nucleotide binding and will therefore be outlined in some detail. [1]H
resonances of many of the aromatic protons overlap, but

Fig. 5 (opposite). The 270-MHz ^{1}H NMR of gene 5 protein, 2 × 10^{-3} M
(A); plus 1 equiv. of d(pT) (B); A-B is the difference spectrum plotted by setting the C(2)H histidyl resonance (peak 1) and the lysyl
ε-CH$_2$ resonance at 3.1 ppm at the same amplitudes in both spectra.
(C) The 270-MHz ^{1}H NMR of d(pT) , 3 × 10^{-3} M. Primed and unprimed
numbers will refer to the carbon atom carrying nonexchangeable protons in the sugar and base, respectively (from reference 14).

major peaks occurring in the spectra of the protein, the
complex with d(pT)$_8$ and the difference spectrum are well-
resolved and have been numbered 1 to 10 in Fig. 5. Their
chemical shifts and proposed assignments are given in
Table 2. On formation of the octanucleotide complex, peaks
3, 4 and 9 disappear from the aromatic region of the protein spectrum, while peak 1 (the C(2)H of His 64) remains
unchanged as do peaks 5, 8 and most of peak 7. In spectra
of the protein at lower concentration, peak 8 is clearly
resolved and corresponds to one proton and has thus been

TABLE 2

*Chemical Shifts and Proposed Assignments of Aromatic Proton Resonances
in Gene 5 Protein and its Complex with d(pT)$_8$*

Peak Number	Chemical Shift ppm	Proposed Assignment
1	8.02	C(2)H His 64 (protein and complex)
2	7.65 to 7.45	C(6)H Thymiding (complex)
3	7.42	Tyr (protein)
4	7.35	Tyr (protein)
5	7.28	Phe (protein and complex)
6	7.05	Tyr (complex)
7	6.90	Tyr and Phe (protein and complex)
7'	6.90	Tyr (protein)
8	6.72	C(4)H His 64
9	6.49	Tyr (protein)
10	6.33	Tyr (complex)
	Ring Current Shifts in Complex ppm	
3	.37	Tyr,
4	.30	Tyr,
9	.17	Tyr, 26, 41 and 56
7'	~.6	Tyr,

assigned to the C(4)H of His 64. A part of peak 7, 7', disappears in the spectrum of the complex. Peak 2 in the complex, which contains at least two resonances of different
chemical shift, appears to represent a broadened resonance
arising from the C(6)H protons of the thymiding ring. Since
the lines are broadened in the spectrum of the $d(pT)_8$ complex, it is difficult to match exactly resonances dis-
appearing from the spectrum of the free protein with new
resonances appearing in the complex. Qualitatively, however, resonances from peaks 3 and 4 in the spectrum of the
protein would appear to be present near 7 ppm (peak 6) in
the spectrum of the complex, while resonances from peaks 7
and 9 appear in the broad region labelled peak 10 in the complex. It is also possible that resonances from some of the
protons in the complex are broadened beyond detection.

The sum of the areas under peaks 3, 4 and 7' of A—B
(Fig. 5) represents ~35 per cent of the original area under
the aromatic resonances of the protein. Thus the environments of about one third of the aromatic protons of the
protein are altered on nucleotide binding, compatible with
the assignment of these particular resonances to the aromatic protons of the 3 tyrosyl residues (Tyr 26, 41, and 56)
that ar known to interact with the nucleotide. These resonances appear to undergo upfield shifts from 0.17 to
0.6 ppm on nucleotide binding (Table 2).

The proton resonances of the methyl groups of the
branched aliphatic side chains of the protein undergo
broadening as well as significant chemical shifts on octanucleotide binding as represented by a number of positive
and negative peaks in the highfield region of the difference spectrum (A—B, Fig. 5). Which of the negative peaks
represents the protons of the thymidine methyl groups in
the complex is not certain, but they might include one or
both peaks at 2.0 and 1.5 ppm. The resonances of the $\varepsilon\text{-}CH_2$
protons of the lysyl residues at ~3.1 ppm undergo neither
broadening nor significant chemical shift in the complex

and will be examined in more detail below. One of the resonances (that at lower field) attributed to the δ-CH_2 protons of the arginyl residues disappears in the spectrum
of the $d(pT)_8$ complex.

*^{1}H-NMR of Gene 5 Protein-Deoxynucleotide Complexes as a
Function of the Size of the Nucleotide*

The resonances in the highfield region of the 270 MHz
[1]H spectrum of gene 5 protein are shown in Fig. 6 and compared to the spectra of two representative tetranucleotide
complexes. From these spectra it is apparent that the
large changes observed in the envelope of the aliphatic
proton resonances on octanucleotide complex formation
(Fig. 5B) appear to be primarily related to changes in
protein structure occurring on cooperative binding of 2
gene 5 monomers to an octanucleotide, since they are much
less marked in the case of tetranucleotide complex formation (Fig. 6). It is likely that many of the highfield
changes in Fig. 5B represent conformational changes induced by protein-protein interactions between monomers and
may also represent changes in residues trapped in the hydrophobic environment of the monomer-monomer interface.

*^{1}H-NMR of Gene 5 Protein-Nucleotide Complexes (Aromatic
Protons) as a Function of Base Sequence and Composition*

In contrast to the changes in the highfield proton
spectrum of gene 5 protein on deoxynucleotide complex
formation, the changes in the lowfield region representing
the aromatic protons of the 5 tyrosyl, 3 phenylalanyl, and
1 histidyl residue appear independent of the size of the
nucleotide, and depend primarily on the nature of the
bases, purine or pyrimidine and their sequence. This conclusion will be documented in more detail by a series of
difference spectra, but the qualitative nature of the
chemical shifts of the aromatic proton resonances of gene
5 protein induced on deoxynucleotide complex formation are

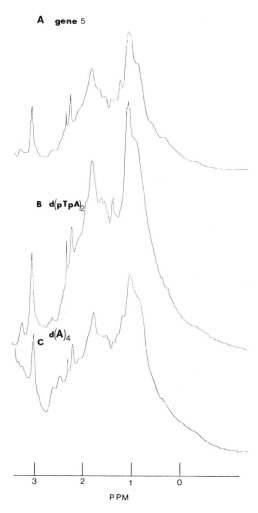

Fig. 6. The 270-MHz NMR in the highfield region of gene 5 protein; 1×10^{-3} M (A); plus 1 equiv. d(pTpA)$_2$ (B); plus 1 equiv. d(pA)$_4$ (C).

best illustrated by the aromatic proton spectra of the complexes of gene 5 protein with three representative tetranucleotides, d(pTpA)$_2$, d(p pT)$_2$ and d(pA)$_4$, overlying the spectra of the protein alone (Fig. 7). The groups of proton resonances of the protein which shift are remarkably similar in each case and involve peaks 3, 4 and 9, resonances also involved in the aromatic shifts induced by

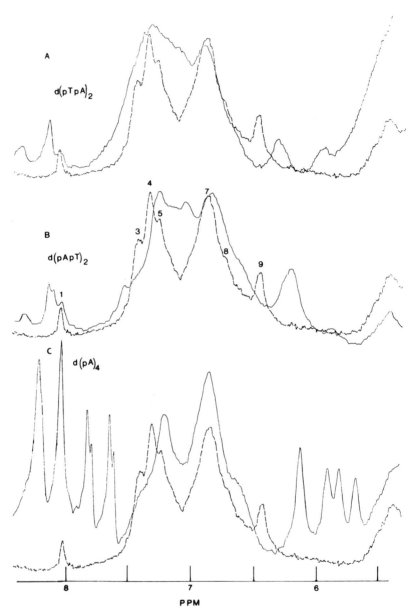

Fig. 7. The 270-MHz NMR in the lowfield region (aromatic protons) of the gene 5 protein-d(pTpA)$_2$ complex (A); of the gene 5 protein-d(pApT)$_2$ complex (B); and of the gene 5 protein-d(pA)$_4$ complex (C). (---) represents the spectrum of the uncomplexed protein. Peak numbers as in Fig. 5.

$d(pT)_8$ (Fig. 5) and assigned to tyrosyl protons (Table 2).

While the resonances of individual tyrosyl protons are not resolved in the gene 5 NMR spectra, perhaps reflecting a smearing out of their chemical shifts arising from the presence of some aggregation of gene 5 monomers, the general upfield shifts of groups of tyrosyl protons are clear (Fig. 7). The shifts induced by $d(pTpA)_2$ and $d(pApT)_2$ are of similar magnitude, resonance from peaks 3 and 4 at 7.42 and 7.35 ppm moving upfield ~0.3 ppm and appearing as several overlapping resonances in the region of 7 ppm (Figs. 7A and 7B). While generally similar, the exact patterns and amount of upfield shift in the two nucleotide complexes are different, suggesting that binding of an isolated tetranucleotide to gene 5 protein is not in a random orientation, but probably maintains a strict 3'-5' orientation relative to the protein surface. Thus an alternating A—T or T—A sequence is differentiated. This effect could be more sharply delineated with the sequence pApApTpT and its inverse and is the object of further study in conjunction with attempts to shift separately the 3 surface fluorotyrosyl resonances with similar specific alterations in base sequence between pyrimidines and purines.

Purines in tetranucleotides, while causing upfield shifts of the same group of tyrosyl protons (peaks 3, 4 and 9), clearly induce greater upfield shifts as indicated by the full aromatic proton spectrum of the $d(pA)_4$ complex (Fig. 7C). A new resonance now appears in the complex well upfield of 7 ppm and overlapping the original peak 7. This difference in the relative magnitude of the upfield shifts induced by pyrimidine and purine bases is even more graphically illustrated by complex formation with the octanucleotide pair, $d(pT)_8$ and $d(pA)_8$. The aromatic proton shifts induced by binding of gene 5 protein to these nucleotides are presented in Fig. 8 as difference spectra (spectrum of the protein - spectrum of the complex). The

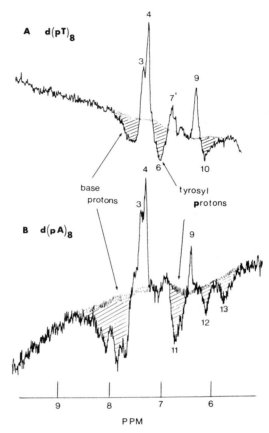

Fig. 8. The 270-MHz NMR difference spectra (protein-complex) for the octanucleotide complexes of gene 5 protein; (A) d(pT)$_8$; (B) d(pA)$_8$.

aromatic peaks that disappear from the protein on complex formation are identical in both cases and are given by the peaks above the baseline. New resonances appearing in the complex are given by the hatched peaks below the baseline. In the d(pT)$_8$ complex the first new resonance appears ~0.3 ppm upfield from peaks 3 and 4 (peak 6), while in the d(pA)$_8$ complex the first new resonance is 0.7 to 0.8 ppm upfield (peak 11). These relative upfield shifts of the tyrosyl protons induced by T and A are in agreement with the calculations of expected ring current shifts for sin- gle protons placed at van der Waals contact distances over

pyrimidine and purine bases respectively [25,26]. The most
upfield tyrosyl protons of the uncomplexed protein, peak
9, disappears in all complexes (Figs. 7 and 8). This resonance appears to move upfield at least 0.2 to 0.3 ppm and
contributes to the group of resonances arising from the 1'
protons of the sugar residues (e.g. peaks 12 and 13, Fig.
8).

Aromatic proton difference spectra for the several
tetranucleotide complexes (Fig. 9) are similar to those
observed for the octanucleotide complexes. The same peaks
(3,4 and 9) are observed to disappear from the protein on
complex formation and new resonance appears in the complex
at similar chemical shift. The complex with $d(pA)_4$, like
$d(pA)_8$, shows the largest upfield shift. Thus the aromatic
proton upfield shifts are independent of the length of the
nucleotide suggesting that each sequential 4 base section
of a nucleotide interacts with the binding groove in the
same general fashion.

^{31}P-NMR of Nucleotide Backbone in the Gene 5-Deoxynucleo-tide Complexes

There have recently been advances in the experimental
information and theoretical interpretation of the ^{31}P
chemical shift values for the phosphorous of the 3'–5'
phosphodiester linkage in polynucleotides. Changes in the
chemical shift have been related to coupled alterations in
the O–P–O bond angle, ϑ, and the torsional (dihedral)
angles, ω, describing the geometry of the C–O–P–O moiety.
The general conclusions are that conformations in which
significant base stacking occurs (helical structures) show
upfield shifts relative to their unstacked conformations.
There appears to be little doubt that the phosphate ester
torsional conformation is gauche-gauche, g-g, in the helical structures [27] and these conformations show the greatest upfield shifts, 0.5 to 1.2 ppm, relative to that for
85 per cent phosphoric acid. In support of this argument,

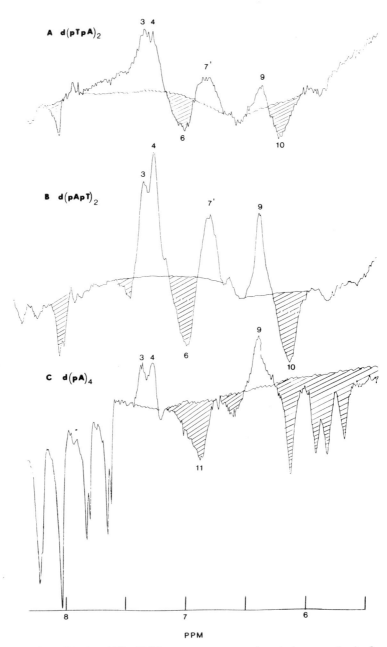

Fig. 9. The 270-MHz NMR difference spectra (protein-complex) for the tetranucleotide complexes of gene 5 protein; (A) d(pTpA)$_2$; (B) d(pApT)$_2$; (C) d(pA)$_4$.

melting of helical forms presumably leading to an increase
in the population of gauche-trans, g-t, conformations are
consistently associated with downfield chemical shifts of
the ^{31}P resonances of the phosphodiesters [27]. The ^{31}P-
NMR spectra of the two octanucleotide complexes of gene 5
protein are shown in Fig. 10 and compared to the spectrum
of free d(pA)$_8$. The 3'-5' diester resonance moves upfield
in both complexes, but the upfield shift is much more
marked for the tight (K_d = 10^{-9} M) d(pA)$_8$ complex, ~1 ppm,
than for the more loosely bound d(pT)$_8$. Both findings suggest that the diester conformation in the complex may be
more like that found in helical conformations (g-g conformation). The difference in the magnitude of the upfield
shift of the diester resonance in the two complexes must
reflect the difference in the stability of the two complexes, the tighter binding of d(pA)$_8$ producing a greater
decrease in the torsional angles. The slight differences
in environment for two groups of diester bonds in the
d(pT)$_8$ complex may represent smaller conformational distortions, consistent with the reduced association constant
of the complex.

The maintenance of a phosphodiester backbone conformation in a single stranded DNA-protein complex which is
similar to that found in base-stacked polynucleotides
could have significant functional implications. Subsequent
transformation of these complexes either by enzymes, poly-
merases, or other packaging proteins might be facilitated
by the presence of the helical conformation in the single
strand.

Model Building

We have previously presented a model of the secondary
structure of the gene 5 protein based on the prediction
of the secondary folding from the amino acid sequence by
the method of Chou and Fasman [10,29]. This secondary
structure can be folded into a three-dimensional model by

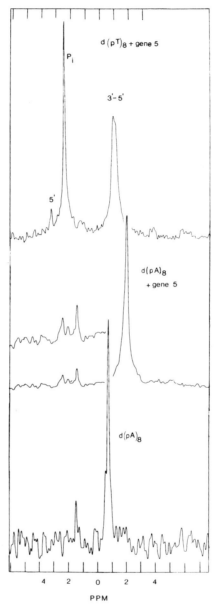

Fig. 10. The ^{31}P-NMR spectra of the gene 5 protein-d(pT)$_8$ complex (A); pH 7.0; the gene 5 protein-d(pA)$_8$ complex (B), pH 8.0; and d(pA)$_8$ (C), pH 7.0.

attempting to provide for as many as possible of the sus-
pected interactions between specific residues of the pro-

tein and the nucleotide based on the NMR and chemical
modification data [14]. The specific modelling of the DNA
binding groove is emphasised in Fig. 11, a stereoview of

Fig. 11. Stereoview of a 'Quick Fit' model of the gene 5 protein-
d(pT)$_4$ complex.

a 'quick fit' model of the protein residues in the binding
groove of the protein with a tetranucleotide, d(pT)$_4$,
about to interact with this groove. The 3 surface tyrosyls, 26, 56, and 41 have been arranged in a stack to
intercalate with the bases of the nucleotide. The precise
stacking arrangement is arbitrary in that alternate stacking can be accommodated by altering the exact positions of
tyrosyls 26 and 56. Tyrosyl 41 at the end of a long region
of β-sheet is relatively fixed in this model. Cys 33 is
placed at the base of the groove to be covered by the
nucleotide. His 64 is at the rear of the model. No evidence has been found for any interaction of this residue
with the nucleotide. The two buried tyrosyls, 34 and 61,
appear behind the DNA binding groove between the two major
regions of pleated sheet.

A maximum number of lysyl residues have been brought
near the nucleotide binding groove (e.g. Lys 3, 7 and 24
on the left, Fig. 11) following the generally held view
that most DNA binding proteins, classically histones,

possess groups of positively charged residues, lysines or
arginines, which function by neutralising the negatively
charged phosphate backbone. In the case of gene 5 protein
such a conclusion is supported by the acetylation data
[10]. Interactions of this kind are generally pictured as
the formation of specific salt bridges (hydrogen bonds)
between the $-NH_3^+$ group and the $^-O-PO_3$ group. However, the
^1H resonances of the lysyl $\varepsilon-CH_2$ groups undergo no chemical shift in any of the nucleotide complexes (Figs. 6 and
7). Furthermore Carr-Purcell A pulse sequences carried out
on the protein and its octanucleotide complexes show that
there is practically no change in T_2 values of the $\varepsilon-CH_2$
protons on complex formation [14]. These findings suggest
that the lysyl interaction might better be viewed as a
neutralizing charge cloud rather than as the formation of
a rigid salt-bridged or hydrogen bonded structure. Such a
model might be attractive in the case of a protein which
must go on and off the DNA with facility, but maintain a
rigid structure of low dissociation constant in the complex. The 'cog wheel' base-tyrosyl intercalation at the
van der Waals contact distance of 3.4 Å would provide a
rigid structure, while the lysine charge cloud would provide electrostatic stabilisation energy without introduc-
ing a kinetic barrier to dissociation of the complex at
the cytoplasmic surface of the cell membrane where gene 5
is replaced by gene 8 protein as the DNA is extruded
through the membrane [5].

Determination of significance of the above mechanism
as a general one for protein-nucleic acid interactions
will require detailed studies of more DNA binding proteins. It is of interest that nitration of 5 of the 9 tyrosyl residues in gene 32 protein (the melting protein from
T4 bacteriophage assisting in phage DNA polymerase action)
results in loss of DNA binding [30]. These tyrosyls are
protected from nitration by prior DNA binding [30]. Nitration of 3 to 4 of the 16 tyrosyl residues in T7 RNA poly-

merase also abolishes binding of that enzyme to the pro-
moter [30]. It might also be speculated that the 4 tyrosyl
residues (7, 12, 17 and 47) in the DNA-binding N-terminal
amino acid sequence of the lac repressor (residues 1 to
59 [31]) also participate in binding to the DNA.

Acknowledgements

This work was supported by Grants No. GM 21919-02 and
AM 18778 from the National Institutes of Health and by
Grant No. 1-PO7-PROO798 from the Division of Health Re-
sources, National Institutes of Health. Initial 270 MHz
proton NMR studies were done at Oxford University while
one of us, J.E.C., was on leave in the laboratory of Professor R.J.P. Williams. We gratefully acknowledge the many
contributions of members of the Oxford Enzyme Group. The
helpful discussions of Dr. Jan Chlebowski are gratefully
acknowledged. We thank Ms. Judy Pascale-Judd for excellent
technical assistance.

References

1. Salstrom, J.S. and Pratt, D. (1971). *J. Mol. Biol.* **61**, 489–501.
2. Alberts, B., Frey, L. and Delius, H. (1972). *J. Mol. Biol.* **68**, 139–52.
3. Mazur, B.J. and Model, P. (1973). *J. Mol. Biol.* **78**, 285–300.
4. Marvin, D.A. and Hohn, B. (1969). *Bact. Rev.* **33**, 172–209.
5. Kornberg, A. (1974). *In: DNA Synthesis*, W.H. Freeman and Co., San Francisco, p. 242.
6. Pratt, D., Laws, P. and Griffith, J. (1974). *J. Mol. Biol.* **82**, 425–39.
7. Berkowitz, S.A. and Day, L.A. (1974). *Biochemistry* **13**, 4825–31.
8. Nakashima, Y., Dunker, A.K., Marvin, D.A. and Konigsberg, W. (1974). *FEBS Lett.* **43**, 125.
9. Cuypers, T., van Ouderaa, F.J. and de Jong, W.W. (1974). *Biochem. Biophys. Res.* **59**, 557.
10. Anderson, R.A., Nakashima, Y. and Coleman, J.E. (1975). *Biochemistry* **14**, 907–17.
11. Day, L.A. (1973). *Biochemistry* **12**, 5329–39.
12. Hoult, D.I. and Richards, R.E. (1975). *Proc. R. Soc. London, Ser. A.* 311–40.
13. Chlebowski, J.F., Armitage, I.M., Tusa, P.P. and Coleman, J.E. (1976). *J. Biol. Chem.* **251**, 1207–1216.
14. Coleman, J.E., Anderson, R.A., Ratcliffe, R.G. and Armitage, I.M. (1976). *Biochemistry* **15**, 5410–30.
15. Nakashima, Y. and Konigsberg, W. (1975). International Symposium

for Photobiology, Williamsburg, Va.

16. Ponzy, Lu, Jarema, M., Mosser, K. and Daniel, W.E. (1976). *Proc. Nat. Acad. Sci. U.S.* **73**, 3471–5.
17. Hull, W.E. and Sykes, B.D. (1974). *Biochemistry* **13**, 3431–7.
18. Hull, W.E. and Sykes, B.D. (1975). *J. Mol. Biol.* **98**, 121–53.
19. Hull, W.E. and Sykes, B.D. (1975). *Biochemistry* **15**, 1535–46.
20. Roberts, G.C.K. and Jardetzky, O. (1970). *Adv. Protein Chem.* **24**, 447–545.
21. Dwek, R.A. (1973). *Nuclear Magnetic Resonance in Biological Systems: Applications to Enzyme Systems*, Clarendon Press, Oxford, p. 76.
22. Campbell, I.D., Lindskog, S. and White, A.I. (1974). *J. Mol. Biol.* **90**, 469–89.
23. Dobson, C.D., Moore, G.R. and Williams, R.J.P. (1975). *FEBS Lett.* **51**, 60–5.
24. Bradbury, E.M., Cary, P.D., Crane-Robinson, C. and Rattle, H.W.E. (1973). *Ann. N.Y. Acad. Sci.* **222**, 266–89.
25. Giessner-Prettre, C. and Pullman, B. (1970). *J. Theor. Biol.* **27**, 87–95.
26. Giessner-Prettre, C. and Pullman, B. (1971). *J. Theor. Biol.* **31**, 287–97.
27. Gorenstein, D.G., Findlay, J.B., Momtï, R.K., Luxon, B.A. and Kar, D. (1976). *Biochemistry* **15**, 3796–803.
28. Patel, D.J. (1974). *Biochemistry* **13**, 2388–96.
29. Chou, P.Y., Adler, A.J. and Fasman, G.D. (1975). *J. Mol. Biol.* **96**, 29.
30. Anderson, R.A. and Coleman, J.E. (1975). *Biochemistry* **14**, 5485–91.
31. Geisler, N. and Weber, K. (1977). *Biochemistry* **16**, 938–43.

NUCLEIC ACID NMR STRUCTURAL STUDIES

G.T. ROBILLARD

Department of Physical Chemistry, University of Groningen
Zernikelaan, Groningen, The Netherlands

Introduction

In 1971 proton resonances from tRNA in H_2O were found in
the region of -15 to -10 ppm and attributed to ring-N-H...
N-ring hydrogen bond interactions from the Watson-Crick
base pairs in the cloverleaf structure [1]. Since the number of resonances seemed to coincide roughly with the
number of proposed Watson-Crick interactions in the secondary structure, ring current shift prediction methods
were derived as a first stage in assigning spectra [17].
Combined with the elegant fragment assignment procedures
of Lightfoot and co-workers [2], there appeared to be a
prosperous future awaiting NMR in nucleic acid physical
chemical studies. Such convictions were only strengthened
when Crothers *et al.* [3] were able to map optical melting
and T-jump data onto tRNA structural transitions monitored
by proton NMR and thus correlate kinetic parameters with
NMR measurements.

With the solution of the three dimensional crystal
structure came the information that, in addition to those
from secondary structure, Watson-Crick interactions were
also involved in stabilizing tertiary structure. The subsequent investigations of Reid and coworkers proved that,
indeed, additional resonances were present in the low
field spectral region which could account for the crystal-

lographically observed tertiary structure interactions
[4—7]. From these and other studies proof has been provided that at least two of the crystallographically observed
tertiary structure interactions, U8—A14 and G22—m[7]G46, do
yield resonances in the low field spectral region [4,8,9].
Thus there is NMR support for the notion that the tertiary
structure in the crystal and in solution has at least some
common features. With this notion in mind we decided to
attempt to calculate the NMR spectrum of a tRNA from the
X-ray crystal structure atomic coordinates using ring current calculations and a number of admittedly simplified
assumptions. What follows is a summary of the results which
have substantially improved our ability to assign and interpret the low field region of nucleic acid spectra and effectively use NMR to study nucleic acid structure.

Integration

Fig. 1 presents a 360 MHz proton NMR low field spectrum
of yeast tRNA[phe] in H_2O. The numbers listed above the peaks
are integrated intensities for each region which have been
determined by computer using a program which subtracts
Lorentzian lines sequentially from the spectrum. With the
height and width of the Lorentzian line set equal to any
one of the resolved resonances of unit proton intensity in
the spectrum, 26 proton resonances were found in the region
between -15 and -11.5 ppm. Since, in the cloverleaf secondary structure (inset, fig. 1), there are only 20 Watson-
Crick base pairs which should give rise to 20 resonances,
the extra six resonances are suggested to originate in the
tertiary structure interactions shown in the inset [4,5].

Ring Current Calculations

In attempting to calculate a spectrum from the atomic
coordinates of the X-ray crystal structure [13] we have
made the simplifying assumption that the dominant mechanism
determining resonance positions are the ring currents from

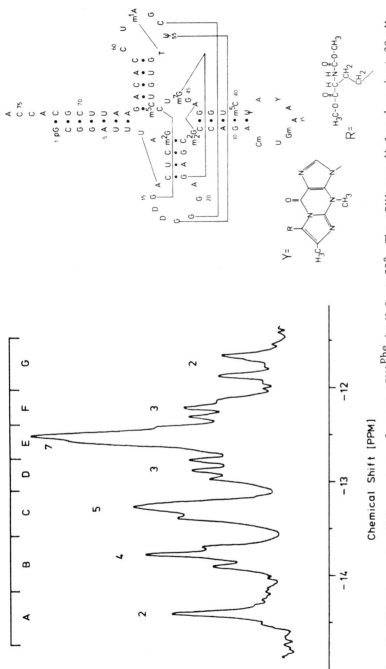

Fig. 1. 360 MHz proton NMR spectrum of yeast tRNA^{Phe} in H$_2$O at 35°. The tRNA was dialyzed against 20 mM KH$_2$PO$_4$, 5 × 10^{-4} M MgCl$_2$ pH 6.8 for 48 hours. *Inset:* Cloverleaf structure of yeast tRNA^{Phe} [10,11]. The black lines indicate the commonly agreed upon tertiary structure interactions which should generate resonances in the low field spectral region [12].

the 76 nucleotide rings in the molecule. Thus, other
effects such as diamagnetic anisotropies and electric
fields whose magnitudes should be substantially less than
the ring current shifts have been neglected for the time
being. The ring current shifts have been calculated by the
procedure of Haigh and Mallion [14] in which all contributions to the ring current shifts, σ_n, are standard geo-
metric terms, K_i, that are calculated from the positions
of the ring atoms and multiplied by a factor, C, proportional to the magnitude of the ring current, J_i, as shown
in Fig. 2. Haigh and Mallion used an empirical factor that

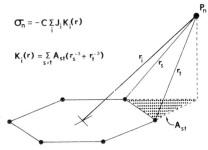

$$\sigma_n = -C \sum_i J_i K_i(r)$$

$$K_i(r) = \sum_{s \neq t} A_{st}(r_s^{-3} + r_t^{-3})$$

Fig. 2. Schematic representation of the Haigh and Mallion ring current calculation procedure [14].

is relevant only for in-plane protons attached to aromatic
ring systems, but that should be scaled up by a factor of
2.6 for large distances from the ring. We have used this
factor of 2.6 as appropriate for the distances occurring
in tRNA.

Two sets of parameters necessary as input for the spectral calculations are 1) the magnitude of the individual
ring currents and 2) the original positions of the hydrogen bond-proton resonances (low field offsets) in the absence of shielding ring currents from neighbouring bases
in the helical stack. Since there are several different
sets of published values for each of these input parameters and no compelling argument for using any one set,
we used the calculated ring currents of Giessner-Prettre
and Pullman [15] and the low field offsets of Shulman *et*

al. [17] only as starting values in an iterative procedure
that successively refined each value. First the A and G
ring currents and the difference between the AU and GC
low field offsets were varied while the C and U values
were held fixed and the computer chose the values which
gave the best fit between the nearest resonances of the
calculated and measured spectra. Subsequently the C and
U ring currents were varied while the new A and G values
were held constant. These new C and U values then served
as the basis for a final slight refinement of the A and G
ring currents. Because of the rapid convergence of this
technique, no further iteration was required. The results
of this procedure are the ring currents presented in Table
1 and the calculated spectrum in Fig. 3B [16]. As can be

TABLE 1

Ring Current Values [1]

Nucleotide	Ring Current	
	Giessner-Prettre & Pullman [15]	Optimized
Adenine		
Hexagonal ring	0.88	0.76
Pentagonal ring	0.67	0.58
Guanine		
Hexagonal ring	0.25	0.29
Pentagonal ring	0.63	0.72
Cytosine	2.27	0.21
Uracil	0.08	0.11

seen in Table 1, the optimized ring currents are somewhat
different from those calculated by Giessner-Prettre and
Pullman [15]. However, the direction as well as the magnitude of the changes are what should be anticipated as a
result of hydrogen bonding. Giessner-Prettre and Pullman

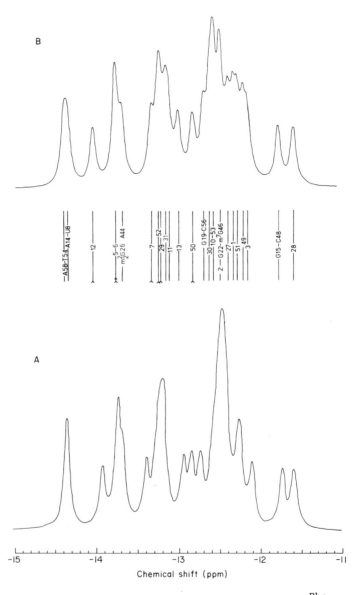

Fig. 3. Spectrum A: Computer simulation of a yeast tRNA^Phe NMR spectrum. Spectrum B: Ring current calculated spectrum based on optimized ring currents, Table 1, and atomic coordinates from the X-ray crystal structure [13]. The indicated base pair positions are those assigned by the calculations.

calculated values for free bases while our optimization
was performed on a hydrogen bonded system. During hydrogen
bonding the electron density and hence the ring current
should increase for the proton donor base and decrease
for the acceptor base precisely as is observed in Table 1.

The resulting calculated spectrum in Fig. 3B shows a
striking resemblance to the computer simulation of the
measured yeast tRNA[Phe] spectrum in Fig. 3A. In part these
results are due to a new low field offset chosen for the
AU Watson-Crick hydrogen bond proton resonances. Before
the three dimensional structure of tRNA was known -14.8
ppm was chosen as the low field AU offset in order to account for resonances observed below -14.0 ppm [17]. Realiz-
ing now that two reversed Hoogstein AU interactions occur
in the tertiary structure, one of which has already been
experimentally assigned to this low field region [4], we
have used the computer selected Watson-Crick AU offset of
-14.35 ppm and the reversed Hoogstein offset of -14.9 ppm.
The remainder of the non-Watson-Crick resonances have been
assigned on the basis of what experimental evidence already
existed [2,4,8] and, beyond that, on the basis of where
vacancies occurred in the calculated spectrum of the secondary structure resonances [16]. Other than the positioning of the tertiary resonances and the Aψ 31 resonance,
the remainder are simply computer assignments and are not
based on experimental evidence.

The selection of a new AU offset substantially alters
predicted resonance assignments, therefore, it is necessary to determine whether the optimized ring currents and
low field offsets are more suitable than previously used
values in terms of predicting and interpreting low field
NMR spectra. Verification could be obtained either by 1)
calculating the NMR spectrum of a different species of
tRNA from the atomic coordinates of this second species
and/or 2) determining experimentally that the computer
assignments of the yeast tRNA[Phe] spectrum are correct (i.e.

that they explain the experimentally observed temperature
dependence). Both approaches have been successfully investigated and are reported below.

Calculation of the E. Coli tRNA$_1^{Val}$ NMR Spectrum

The cloverleaf sequences of yeast tRNA^{Phe} and E. Coli
tRNA$_1^{Val}$ are presented in Fig. 4. Each molecule contains 76
bases and each corresponding loop and helix contains the

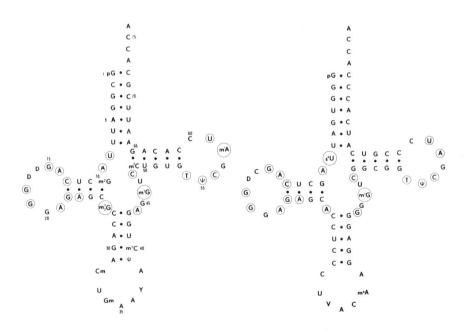

Fig. 4. The cloverleaf sequences of yeast tRNA^{Phe} [10,11] and E. Coli
tRNA$_1^{Val}$ [18-21]. The encircled bases are those involved in tertiary
structure interactions [22-24].

same number of bases. The encircled bases in the yeast
tRNA^{Phe} cloverleaf are those which participate in crystallographically determined tertiary structure interactions
[22-24]. Comparison of these bases with the encircled
bases in the cloverleaf of E. Coli tRNA$_1^{Val}$ shows that
every base in yeast tRNA^{Phe} participating in tertiary
structure interactions corresponds to an identical base
in E. Coli tRNA$_1^{Val}$ except for the switch between G and A
in positions 26 and 44. These similarities suggest that
the two structures should be similar if not identical.
Therefore, since no other tRNA structure has been solved
crystallographically, we built a structure of E. Coli
tRNA$_1^{Val}$ from that of yeast tRNA^{Phe} by the following procedure. Holding the atomic coordinates of the phosphate-
ribose backbone of yeast tRNA^{Phe} fixed, tRNA$_1^{Val}$ base substitutions were made, by computer, at each point where the
two structures differed. After the base substitutions were
completed the atomic coordinates of the new tRNA$_1^{Val}$ structure were idealized by minimizing unfavourable stereo-
chemistry and Van der Waals interactions. The resulting
coordinates, considered to represent the three dimensional structure of E. Coli tRNA$_1^{Val}$, were used as input for
calculations of the E. Coli tRNA$_1^{Val}$ NMR spectrum. Since
the primary goal of this experiment was to check the validity of the optimized ring currents and low field offsets used in calculating the yeast tRNA^{Phe} spectrum in
Fig. 3, these same values were used together with the new
atomic coordinates to calculate the NMR spectrum of E.
Coli tRNA$_1^{Val}$.

The results, presented in Fig. 5 [25] compare the measured 360 MHz proton NMR spectrum of E. Coli tRNA$_1^{Val}$ with
the calculated spectrum. It has been shown, previously,
that the measured spectrum contains the same number of
proton resonances in the low field region as that of yeast
tRNA^{Phe} arising, presumably, from the same secondary and
tertiary structure hydrogen bonds. Considering the mani-

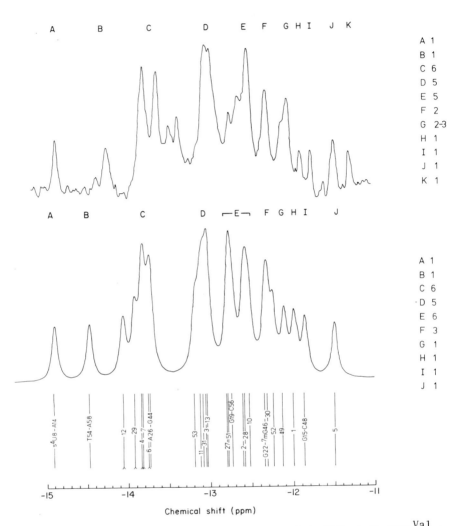

Fig. 5. Upper: 360 MHz proton NMR spectrum of 1 mM E. Coli tRNA$_1^{Val}$ in H$_2$O containing 15 mM MgCl$_2$, 10 mM sodium cacodylate, 1 mM EDTA and 0.1 M NaCl, pH 7. The sample temperature was 45°. Lower: Ring current calculated spectrum using E. Coli tRNA$_1^{Val}$ coordinates and computational procedures described in the text. Secondary structure AU resonances are marked (∧) at the bottom of the assignment line [25].

pulations and approximations inherent in generating the co-ordinates of E. Coli tRNA$_1^{Val}$, the agreement between the measured and calculated spectra is remarkable. The RMS error between measured and calculated resonance positions is 0.1 ppm. Certainly discrepancies still exist and are clearly visible in the chemical shift positions as well as the integrated intensities listed to the right of each spectrum. However, the general similarity between the spectra is a strong argument in favour of both the optimized ring currents and, more importantly, the modified low field offsets.

Experimental Assignments

The commonly accepted method of verifying tRNA hydrogen bond resonance assignments, or of making those assignments in the first place, is to monitor the temperature dependence of the unfolding process by NMR. As helices begin to open the hydrogen bond proton resonances disappear by the loss of intensity with or without broadening depending on the kinetics of the event [3]. Thus, the opening of a given helix is correlated with the simultaneous loss of a number of resonances. Several apparently successful studies of this type have been reported [3,26,27]. In the case of yeast tRNA^{Phe}, however, no complete analysis of the NMR 'melting' has succeeded. Presumably this is because of the confusion generated by overlapping NMR 'melting' transitions and incorrect spectral assignments. Fortunately, an extensive series of T-jump optical and fluorescence studies have provided a clear picture of the thermal denaturation of yeast tRNA^{Phe} [28-32]. Mg^{++} stabilizes the molecule to the extent that no thermal transitions are observed until 70° at which point the whole molecule unfolds in one cooperative transition. However, in 30 mM Na^{+} pH 6.8, Mg^{++} free yeast tRNA^{Phe} undergoes five distinct melting transitions between 20° and 70°. Table 2 lists these transitions along with their T_M's and relaxation times. Since the optical experiments achieved very good resolution of these transitions in the low Na^{+}, no Mg^{++} buffer we have chosen the same solvent conditions to monitor the NMR 'melting'.

G.T. ROBILLARD

TABLE 2

Transitions observed during the Thermal Denaturation of Yeast tRNA^{Phe} in 30 mM Na^{+}, no Mg++, pH 6.8

Transition	Structure Involved	T$_M$	Relaxation Time	Ref.
1	Tertiary Structure	25°	Slow: 10 msec	[29]
2 and 3	Residual Tertiary structure and the acceptor helix and the anticodon helix	35°–40°	Slow: 2-23 msec and 17-475 msec	[32]
4	TψC Helix	45°–50°	Fast: 20–100 μsec	[28]
5	DHU Helix	60°–65°	Fast: 20–100 μsec	[28]

The plan followed in these experiments was first to demonstrate that, at the tRNA concentrations of the NMR experiment, we were observing the same transitions as seen in the optical and T-jump experiments where the concentrations were 10^2 to 10^4 times lower. This has been demonstrated by NMR studies of the methyl resonance spectral region reported below. Having proved that the same transitions occurred in the NMR samples, we then used the optical, T-jump and high field NMR studies to predict the transitions which should be observed in the hydrogen bond proton resonance spectra as well as the T$_M$'s for these transitions. The resonance losses seen in those transitions were then compared with the predicted resonance positions for the portion of the molecule undergoing the transition.

Transitions Reported by High Field NMR Spectral Changes

In T-jump optical and fluorescence studies base stacking is measured as changes in hyperchromicity or fluorescence intensity. This same stacking can be measured with NMR by following the shift of the methyl proton resonances from the modified bases. In the case of yeast tRNA^{Phe} this

is particularly advantageous since there are a large
number of methylated bases rather evenly distributed
throughout the molecule, including those regions participating in tertiary structure interactions (see Fig. 4).

Fig. 6 shows the temperature dependence of the methyl
proton region of the NMR spectrum of yeast tRNA^{Phe} in the
presence of Mg^{++}. As anticipated from the optical studies,
the molecule is very stable and retains its structure up
to 65°. Between 65° and 75° the NMR spectrum is substantially altered reflecting large structural changes in

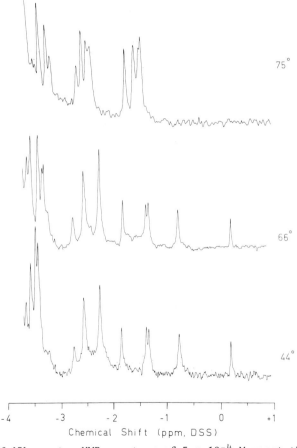

Fig. 6. 360 MHz proton NMR spectrum of 5 × 10^{-4} M yeast tRNA^{Phe} at
various temperatures. The sample was dialyzed against 15 mM MgCl ,
0.1 M NaCl in D$_2$O pH 7 (meter reading) for 48 hours [33].

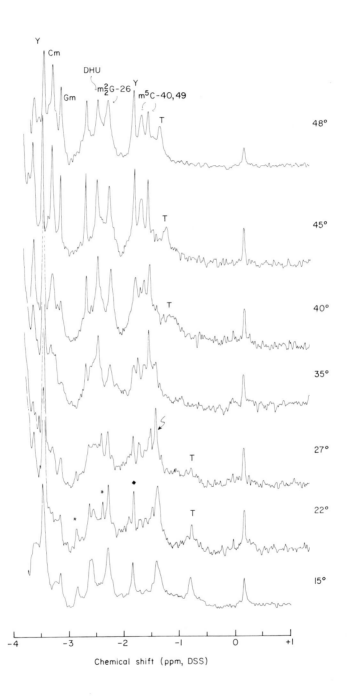

the molecule.

The temperature dependence in 30 mM Na^+ pH 6.8 for Mg^{++} free yeast $tRNA^{Phe}$ is presented in Fig. 7. The similarity of the lowest temperature spectrum in this figure and in Fig. 6 suggests that, at low temperature, the conformation in the absence of Mg^{++} is similar to that in the presence of Mg^{++}. Nevertheless, the structure is less stable in the absence of Mg^{++} and, under these conditions, undergoes several transitions between 20° and 70°. The lowest temperature transition, transition I, in Fig. 7A occurs between 20° and 40°. It is reported by the disappearance of a number of resonances between 0 and -3.7 ppm and the subsequent appearance of others. Transition II can be seen in Figs. 7 and 8 as the shifting of the T—54 methyl resonance between 40° and 60°. Transition III takes place between 50° and 80° as seen in Fig. 7B. After 80° the molecule appears to be in the random coil form. The shifts of the individual methyl resonances report transitions occurring in specific regions of the molecule and can be directly compared with the transitions found by T-jump optical and fluorescence studies listed in Table 2. We will, however, consider these transitions in the reverse order beginning with the high temperature spectra since the assignments are more certain at these temperatures.

Transition III 50° to 80°C

As the temperature is lowered from 80° sturctural reorganization is first reported by slight upfield shifts of the methylene resonances of dihydrouridine 16 and 17 and a large upfield shift of the methyl resonances of

---

Fig. 7A (opposite). 360 MHz proton NMR spectra of Mg^{++} free yeast $tRNA^{Phe}$ (2×10^{-4} M) in 10 mM Na_2DPO_4, 20 mM NaCl, pH 6.8 (with phosphoric acid), 15° to 48°. The 15° and 48° spectra have been signal averaged longer than the rest for a better signal to noise ratio. At the tRNA concentrations used here, the observed transitions are completely reversible [33]. The assignments in the 48° spectra are from Kan *et al.* [34].

216

G.T. ROBILLARD

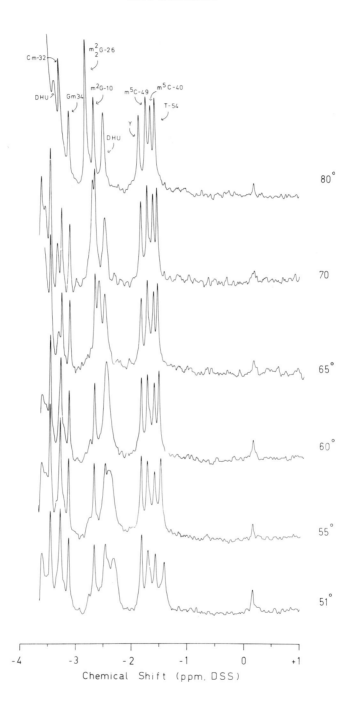

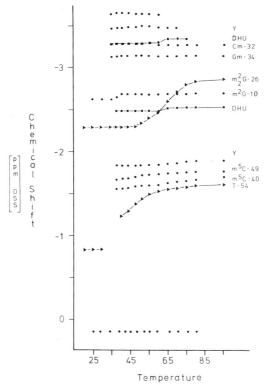

Fig. 8. Plot of the chemical shift as a function of temperature for the resonance positions in Fig. 7 [33].

$m_2^2 G$ 26 (see Figs. 7B and 8). These shifts all occur with the same T_M of 60° to 65°. In Table 2, transition 5 observed by T-jump studies on intact yeast tRNA[Phe] as well as the isolated 5'½ fragment of this molecule, occurs with a T_M of 60° to 65° and has been interpreted as the melting of the DHU helix. This is in agreement with the transition observed by NMR. As the DHU helix forms, shifts are observed for the dihydrouridine methylenes which are in the loop region and the $m_2^2 G$ 26 at the other end of the DHU helix.

Fig. 7B (opposite). 360 MHz proton NMR spectra of Mg[++] free yeast tRNA[Phe] (2×10^{-4} M) in 10 mM Na_2DPO_4, 20 mM NaCl, pH 6.8 (with phosphoric acid), 51° to 80°. The assignments in the 80° spectra are from Kan *et al.* [34].

Transition II 45° to 55°C

As the temperature continues to decrease, the next distinct transition occurs with a T_M of 50° and is reported
by a 0.37 ppm upfield shift of the T 54 methyl resonance.
Since this transition is not reported by any other methyl
resonance, it suggests that the events being sensed are
simply local changes resulting from the formation of the
TψC helix. In agreement with this observation, the optical studies in Table 2 report that the TψC helix melts
with a T_M of 45° to 50°.

The two high temperature transitions are in the fast
exchange limit on the NMR time scale as seen by the smooth
shifting of resonances. Such behaviour is expected since
the relaxation times of these two transitions reported by
T-jump are in the microsecond range. However, the three
lowest temperature transitions in Table 2 have relaxation
times generally larger than 10 milliseconds meaning that,
for chemical shift differences between two states of more
than a few hertz, intermediate or slow exchange NMR behaviour is expected.

Transition I 20° to 40°C

In the 45° spectrum, Fig. 7A, the thymine resonance
begins to decrease in intensity and broaden. Its disappearance signals the end of the fast exchange transition
II and the beginning of another structural change with
intermediate or slow exchange characteristics. The resonance finally reappears in the 27° spectrum shifted upfield
to -0.8 ppm. A similar decrease in intensity is observed
for the resonances from m^5C 40 and m^5C 49 in the 45°
spectrum and an increase of the resonance marked by the
arrow in the 27° spectrum. The Y base resonance at -1.85
ppm undergoes a similar change but, in this case, is
shifted downfield to the position marked by the diamond
in the 22° spectrum. In this same temperature range similar shifts are seen for the resonances marked by asterisks

in the 22° spectrum as well as the resonances for Cm 32 and Gm 34. As can be seen from Fig. 4, the methyl resonances reporting this transition arise from bases participating either in tertiary structure interactions or in the secondary structure of the anticodon stem and loop. The temperature dependent shifts between 20° and 40° suggest that, during this transition, the molecule is undergoing widespread conformational changes involving the DHU, TψC and anticodon loops and helices and the junction between them. Table 2 supports these observations showing that the three lowest temperature transitions occur with T_M's of 25° and 35° (i.e. within the 20° to 40° transition range seen by NMR). They involve the opening of tertiary structure followed by the unwinding of anticodon and acceptor helices. Unfortunately, because of the broad temperature range for the NMR transition I, the individual transitions seen by T-jump cannot be resolved. Nevertheless, it is clear that the same events monitored by T-jump techniques are reported by the methyl resonances.

Transitions reported by the Hydrogen Bond Spectral Region

From the comparisons made in the previous section it is clear that the high tRNA concentrations used in the NMR studies do not alter the sequence of melting transitions or the T_M's of these transitions as determined by optical melting and T-jump techniques. Therefore, we can now predict the sequence of transitions which should be reported by the hydrogen bond proton resonances in the low field spectral region as well as the temperatures at which they should occur. The theoretical background for this type of analysis has been previously published and will not be considered here [3].

The melting of tertiary structure, transition 1 in Table 2, is a slow relaxation process and 'melting' in the low field spectral region should occur by a simple loss of resonance intensity without broadening in the region of

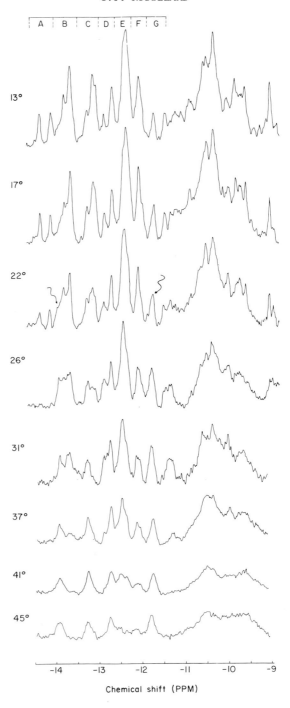

Chemical shift (PPM)

T_M = 25°. Transitions 2 and 3, the melting of the anti-codon and acceptor helices, are also slow relaxation processes which should be reported by a similar loss of intensity without broadening but in the region of T_M = 35°. Transitions 4 and 5, the melting of the TΨC and DHU helices, are fast relaxation processes. When the relaxation times are extrapolated to the NMR time range according to the method of Crothers *et al*. [3] the expectation is that both of these transitions will be reported by a loss of intensity with broadening at temperatures below the optically observed T_M's. The TΨC helix resonances should begin to broaden and disappear at about 40° and the DHU helix resonances at somewhat higher temperature [33].

Fig. 9 presents the temperature dependence of the low field resonances for Mg^{++} free yeast tRNA^{Phe} in 30 mM Na$^+$ pH 6.8 [33]. The only major differences between the low temperature Mg^{++} free spectrum and that of yeast tRNA^{Phe} in the presence of Mg^{++}, Fig. 1, are that the peak in region A has been split into two single resonances and one resonance is missing from the peak in region C. Otherwise, relative to both intensities and positions of resonances, the spectra are quite similar.

Tertiary Structure

The first transition reported in Fig. 9 is the loss of a series of resonances between 13° and 26°. This can be seen more clearly from the computer subtraction in Fig. 10. Both the T-jump and the high field NMR studies tell us that this transition should be the melting of tertiary structure. In agreement, the calculated positions for the tertiary structure hydrogen bond resonances listed at the bottom of Fig. 10 show that for every region in which a tertiary structure resonance is predicted, there is a loss of at least one proton resonance. Particularly noteworthy

Fig. 9 (opposite). 360 MHz proton NMR spectra as a function of temperature for Mg^{++} free yeast tRNA^{Phe} in 10 mM Na$_2$HPO$_4$, 20 mM NaCl, pH 6.8 [33].

is the agreement between the predicted positions of the
two reversed Hoogstein AU tertiaries in region A and the
clear loss of two resonances in the same region. Those
positions are predicted only by the optimized ring currents and low field offsets used in this paper as are the
similar predictions in the E. Coli tRNA$_1^{Val}$ spectrum in
Fig. 5. As mentioned earlier, removal of the m^{7}G from
position 46, chemically, results in the loss of a resonance in region E thereby experimentally assigning the
G22–m^{7}G 46 hydrogen bond proton resonance to the same
region where we had calculated it and where we now observe
a resonance loss [9].

In the region of -9 to -11 ppm, Fig. 10 also shows four
clear resonance losses. It has been demonstrated earlier
that resonances from GU wobble interactions occur in this
region [27]. The losses observed here presumably arise
from tertiary structure interactions involving hydrogen
bonding arrangements similar to GU wobble interactions.
For example G15–C48 and G18–ψ55 and ψ55–PO58 have ring NH
protons bonded to exocyclic oxygens while the A9–A23 and
G15–C48 interactions have exocyclic amino groups directly
bonded to ring nitrogens.

Peak C in Fig. 10 is exceptional since an intensity of
about 2 protons is lost in the first transition while no
tertiary structure resonances are predicted in this region. Part of the intensity loss is related to two resonances which report the first transition not by 'melting'
but by shifting their position in slow exchange. Thus in
Fig. 8 as the intensity on the upfield side of peak B
decreases, intensity on the low field shoulder of the same
peak increases. Similarly, as the intensity of peak C decreases, intensity of peak G increases (see arrows in the
22° spectrum, Fig. 9, and the shaded area of the difference spectrum, Fig. 10). As will be shown later these resonances arise from AU 12 and GC 13 which are reporting the
breaking of the tertiary structure interactions in their

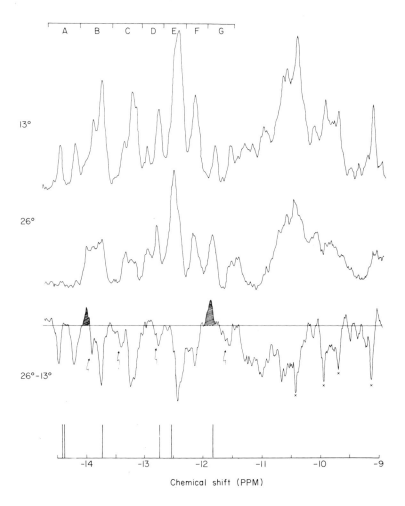

Fig. 10. Computer subtraction of the 13° and 26° spectra in Fig. 9. The lines mark the positions where tertiary structure resonances have been predicted to occur (see Fig. 3) [33].

immediate vicinity.

The GC 13 resonance was calculated to occur in peak C and its shift to peak G accounts for the loss of one resonance in peak C during the first transition. A second proton is lost in peak C during this transition which arises from the Aψ 31 interaction we will consider below.

The arrows in Fig. 10 mark isolated resonances which

have already begun to melt in the second transition. By
their intensity we estimate that the second transition is
approximately 30 per cent complete at this temperature.
Since two resonances in region B one in region C and three
in region E also melt in this second transition, it is
this partial melting which gives rise to the additional
intensity loss in these regions during transition I.

Acceptor and Anticodon Helices

The melting of the acceptor and anticodon helices is
reported both by T-jump data and NMR data from the high
field spectral region as a slowly relaxation transition
with a T_M of 35°. Following the loss of resonances assigned to tertiary structure another series of resonances
disappear by 37°. The difference between the 26° and 37°
spectra is presented in Fig. 11 along with the calculated
resonance positions for the hydrogen bond protons of these
two helices. Keeping in mind that a partial melting in
this transition has already occurred, as indicated in the
previous section, we find in Fig. 11 a loss of 10 resonances. In peak C and D two resonances are lost while three
are predicted to occur in this region. In the first transition however we have already pointed out that one more
resonance disappeared from region C than expected. This
discrepancy presumably originates in the anomalous 'melting' behaviour of the Aψ 31 resonance. In NMR studies on
the isolated anticodon helix fragment [2] one resonance
also melted earlier than the rest and, because of its
instability, was attributed to the Aψ hydrogen bond.

With this reasoning the loss of resonances intensity
during this transition fits well in numbers and reasonably well in position with those resonance positions calculated for the acceptor and anticodon helices. In this
regard AU 5 and AU 6 are particularly noteworthy since
they are predicted to occur in region B only using our
optimized low field offsets. All other previous offsets

place these resonances in region A.

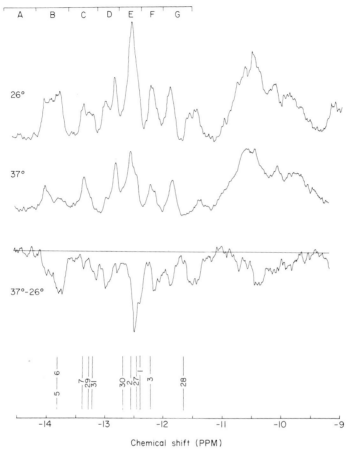

Fig. 11. Computer subtraction of the 26° and 37° spectra from Fig. 9. The indicated positions are the calculated positions for resonances from the anticodon and acceptor helices (Fig. 3) [33].

TψC and DHU Helices

The last two transitions observed by optical, T-jump or high field NMR measurements are rapidly relaxing transitions involving the melting of the TψC and DHU helices at approximately 50° and 60° respectively. As stated earlier, the corresponding hydrogen bonded proton resonances are expected to 'melt' earlier at about 40° and 45° [33]. Fig. 12 shows the difference spectrum resulting from the

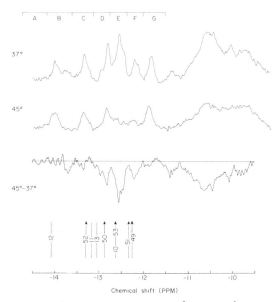

Fig. 12. Computer subtraction of the 37° and 45° spectra from Fig. 9. The indicated positions are the ring current calculated positions for resonances from the TψC and DHU helices. The TψC resonance positions are marked by arrows (see Fig. 3) [33].

subtraction of the 37° and 45° spectra from Fig. 9. In keeping with the predicted behaviour for a rapid relaxation process the resonances broaden and decrease in intensity. A loss of five resonances occurs in this transition. The TψC resonance positions predicted by the calculations are marked by arrows in Fig. 12. Except for the position of AU 52, these losses agree with the predicted resonance positions for the TψC helix hydrogen bonds.

The remaining four resonances at 45° should be, and are, those of the DHU helix. This has been experimentally determined by examining the NMR spectrum of the 5'½ fragment containing only the DHU helix. Spectra of the fragment are identical to the 45° spectrum in Fig. 9 [36]. The assignments in Fig. 12 show that only one predicted position does not correspond with an observed resonance from the DHU helix. We discussed earlier the slow exchange shift of two resonances during the melting of tertiary

structure (see arrows in 22° spectrum, Fig. 9). Clearly
AU 12 is the one which shifts in region B and then remains
in its position through to the 45° spectrum. The resonance
which shifts upfield must be GC 13 which originally occurred in region C where it was calculated. As tertiary structure unfolds, the ring of A 14, previously pulled somewhat
out of the DHU loop stack while bonding to U 8, swings
back into a good helical stack under GC 13 and results in
its shift upfield. Just as the breaking of tertiary structure is a slow relaxation process, so will be the reposi-
tioning of A 14 and therefore, the shift of GC 13 is a
slow exchange shift. Thus in the intact structure GC 13
occurs where it is predicted as do the rest of the resonances for this helix.

It is clear from the analysis presented above that the
predicted low field hydrogen bond proton resonance positions are in good agreement, for the most part, with the
actual resonance positions observed from the melting
studies. We attribute this agreement primarily to the
optimized ring currents and low field offsets used in the
calculations.

Ring Current Shift Estimates

The original ring current shift estimates proposed by
Shulman *et al.* [17] were compiled at the time when -14.8
ppm was being used as a low field offset for AU Watson-
Crick hydrogen bond proton resonances. In order to account
for the distribution of resonances in the low field spectral region while using this offset it was necessary to
increase the A and G ring currents of Giessner-Prettre and
Pullman [15] by 20 per cent. Thus the values presented in
their ring current shift tables [17] are rather high. Because of the success of the modified ring currents and low
field offsets we have calculated new ring current shifts
from idealized A-RNA coordinates using the same ring currents and computational procedures used throughout this

TABLE 3

Ring Current Shifts experienced by Ring NH··N Hydrogen Bonded Protons in A-RNA Helices

5'	U	0.0				A	0.54	(0.11)	3'
	C	0.0				G	0.26	(0.08)	
	G	0.1				C	0.04		
	A	0.25				U	0.02		
				U	A				
				Ո	Ɐ				
	U	0.08				A	0.0		
	C	0.15				G	0.0		
	G	0.39	(0.03)			C	0.0		
3'	A	0.50	(0.05)			U	0.0		5'

5'	U	0.0				A	0.73	(0.1)	3'
	C	0.0				G	0.42	(0.07)	
	G	0.03				C	0.10		
	A	0.09				U	0.05		
				C	G				
				Ɔ	ᓂ				
	U	0.07				A	0.03		
	C	0.13				G	0.0		
	G	0.43	(0.05)			C	0.0		
3'	A	0.65	(0.08)			U	0.0		5'

All values given in ppm

Low Field Offsets
  Watson-Crick AU   -14.35 ppm
  Watson-Crick GC   -13.54 ppm
  Reversed Hoogstein AU   -14.90
  Reversed Hoogstein As_4U -15.40

study. They are presented in Table 3 along with the optimized low field offsets we have been using. It should be emphasized, however, that predictions from such tables will only be reasonably accurate for regions where the RNA helix obeys nearly idealized stacking. The X-ray crystal tructure shows many regions deviating from ideality which will result in observed resonance positions quite different from those estimated from tables.

Conclusions

The calculations presented here are admittedly oversimplified. Such being the case, the results are quite striking. They have allowed us to develop a set of assignments which explain, with virtually no rationalization, the experimental data. In the future, as we eliminate some of the approximations, taking into account some of the small shift contributions, the accuracy and usefulness of such computational procedures is bound to increase.

Acknowledgements

Special credit for this work should be given to B.R. Reid, C.E. Tarr, F. Vosman and H.J.C. Berendsen. Without their help this work would never have been done. I would also like to thank J. Sussman and S.H. Kim for supplying us with coordinates of yeast tRNA^{Phe} and E. Coli tRNA$^{Val}_1$. I am indebted to R. Kaptein and K. Dijkstra for the consistently fine operation of the NMR spectrometer.

The financial support of the ZWO (Netherlands Foundation for Pure Research) for the 360 MHz facility at the University of Groningen is gratefully acknowledged.

References

1. Kearns, D.R., Patel, D., Shulman, R.G. (1971). *Nature* **229**, 338–9.
2. Lightfoot, D.R., Wong, K.L., Kearns, D.R., Reid, B.R. and Shulman, R.G. (1973). *J. Molec. Biol.* **78**, 71–89.
3. Crothers, D.M., Cole, P.E., Hilbers, C.W. and Shulman, R.G. (1974). *J. Molec. Biol.* **87**, 63–88.
4. Reid, B.R., Ribeiro, N.S., Gould, G., Robillard, G.T., Hilbers, C.W. and Shulman, R.G. (1975). *Proc. Nat. Acad. Sci. U.S.A.* **72**, 2049–53.
5. Reid, B.R. and Robillard, G.T. (1975). *Nature* **257**, 287–91.
6. Reid, B.R., Ribeiro, N.S., McCollum, L., Abbate, J., and Hurd, R. (1977). *Biochemistry* (in press).
7. Hurd, R.E., Robillard, G.T., and Reid, B.R. (1977). *Biochemistry* (in press).

8. Daniel, W.E. and Cohn, M. (1975). *Proc. Nat. Acad. Sci. U.S.A.* **72**, 2582–6.
9. Hilbers, C.W. Personal communication.
10. RajBhandary, U.L., Chang, S.H., Stuart, A., Faulkner, R.D., Hoskinson, R.M. and Khorana, H.G. (1967). *Proc. Nat. Acad. Sci. U.S.A.* **57**, 751–8.
11. Nakanishi, K., Furutachi, N., Funamizu, M., Grunberger, D. and Weinstein, I.B. (1970). *J. Am. Chem. Soc.* **92**, 7617–19.
12. Sussman, J.L. and Kim, S.H. (1976). *Science* **192**, 853–8.
13. Sussman, J.L. and Kim, S.H. (1975). *Biochem. Biophys. Res. Comm.* **68**, 89–95.
14. Haigh, C.W. and Mallion, R.B. (1971). *Molec. Phys.* **22**, 955–70.
15. Giessner-Prettre, C. and Pullman, B. (1965). *C.r. hebd. Seanc. Acad. Sci. Paris* **261**, 2521–3.
16. Robillard, G.T., Tarr, C.E., Vosman, F. and Berendsen, H.J.C. (1976). *Nature* **262**, 363–9.
17. Shulman, R.G., Hilbers, C.W., Kearns, D.R., Reid, B.R. and Wong, Y.P. (1973). *J. Molec. Biol.* **78**, 57–69.
18. Yaniv, M. and Barrell, B.G. (1969). *Nature* **222**, 278–9.
19. Harada, F., Kimura, F., Nishimura, S. (1969). *Biochem. Biophys. Acta* **195**, 590–2.
20. Harada, F., Kimura, F. and Nishimura, S. (1971). *Biochem.* **10**, 3269 76.
21. Kimura, F., Harada, F. and Nishimura, S. (1971). *Biochem.* **10**, 3277–83.
22. Kim, S.H., Suddath, F.L., Quigley, G.J., McPherson, A., Sussman, J.L., Wong, A., Seeman, N.C. and Rich, A. (1974). *Science* **185**, 435–40.
23. Sussman, J.L. and Kim, S.H. (1976). *Biochem. Biophys. Res. Comm.* **68**, 89–95.
24. Quigley, G.J., Seeman, N.C., Wong, A.H.J. and Rich, A. (1975). *Nucleic Acid Res.* **2**, 2329–36.
25. Robillard, G.T., Tarr, C.E., Vosman, F. and Sussman, J.L. (1977). *Biophysical Chemistry* (in press).
26. Hilbers, C.W., Robillard, G.T., Shulman, R.G., Blake, R.D., Webb, P.K., Fresco, R. and Reisner, D. (1976). *Biochem.* **15**, 1874–82.
27. Robillard, G.T., Hilbers, C.W., Reid, B.R., Gangloff, J., Dirheimer, G. and Shulman, R.G. (1976). *Biochem.* **15**, 1883–9.
28. Romer, R., Reisner, D., Maass, G., Wintermeyer, W., Thiebe, R. and Zachau, H.G. (1969). *FEBS Lett.* **5**, 1883–9.
29. Romber, R., Reisner, D. and Maass, G. (1970). *FEBS Lett.* **10**, 352–9.
30. Reisner, D., Maass, G., Thiebe, R., Philippsen, P. and Zachau, H. (1973). *Eur. J. Biochem.* **36**, 76–88.
31. Urbanke, C., Romer, R., Maass, G. (1975). *Eur. J. Biochem.* **55**, 439–44.
32. Coutts, S.M., Riesmer, D., Romer, R., Rabl, C.R., Maass, G. (1975). *Biophys. Chem.* **3**, 275–89.
33. Robillard, G.T., Tarr, C.E., Vosman, F. and Reid, B.R. (1977). Submitted for publication.
34. Kan, L.S., Ts'o, P.O.P., Van der Haar, F., Sprinzl, M. and Cramer, F. (1974). *Biochem. Biophys. Res. Comm.* **59**, 22–9.
35. Robillard, G.T. Unpublished data.

MAGNETIC RESONANCE STUDIES OF MODEL AND BIOLOGICAL MEMBRANES

A.C. McLAUGHLIN,* P.R. CULLIS,[†] M. HEMMINGA,[‡]
F.F. BROWN,[§] J. BROCKLEHURST[#]

Introduction

Membranes are one of the most ubiquitous architectural
elements in biological organisms. They may have structural uses, such as providing selectively permeable boun-
daries for cells and subcellular organelles, or they may
themselves be the site of biochemical and physiological
functions, such as mitochondrial electron transport or
nervous conduction.

Membranes are composed mainly of lipids and proteins.
The percentage of protein in the membrane may vary from
as little as 20 per cent for myelin, to as high as 80 per
cent for the inner mitochondrial membrane. The currently
accepted model for the organization of these two structural elements is shown in Fig. 1. There is strong evidence that most of the lipids in biological membranes are
arranged in the bilayer form [1, 2, 3, 4]; i.e. in the
same form as in the aqueous dispersions of purified

*Biology Department, Brookhaven National Laboratories, Upton, L.I.,
N.Y. 11973, U.S.A.
[†]Biochemistry Laboratorium, Rijksuniversiteit Utrecht, Utrecht,
Holland.
[‡]Department of Molecular Physics, Agricultural University,
Wageningen, Holland.
[§]Biochemistry Department, Oxford University, Oxford, U.K.
[#]Imperial Cancer Research Fund, Lincolns Inn Fields, London, U.K.

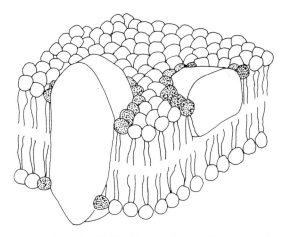

Fig. 1. The current 'simplified' model for the organisation of lipids and proteins in biological membranes. The spheres represent the
polar headgroups of the phospholipid molecules and the wavy lines
projecting from the headgroups represent the fatty acid chains. The
larger globular structures represent protein molecules. The shaded
polar headgroups represent 'boundary-layer' lipid.

lipids. Both the model bilayer membranes and the bilayer
regions of biological membranes exhibit phase transitions
from the rigid 'gel' phase to the more fluid 'liquid-
crystalline' phase as the temperature is raised [1, 5, 6].
These changes in fluidity strongly affect the magnetic
resonance spectra [7, 8] and have been implicated in some
of the physiological functions of the membrane [6, 9, 10,
11]. The proteins are embedded in the bilayer matrix to
varying degrees, with some proteins, such as glycophorin,
traversing the entire membrane [12].

Diffraction techniques (i.e. X-Ray and Neutron) have
provided valuable information on the static structure of
membranes [1, 13, 14, 15, 16]. However, because of their
sensitivity to motion in the 10^{-3} to 10^{-9} second range,
magnetic resonance techniques are particularly well suited for studying the dynamic structure of membranes. EPR
and NMR techniques have provided estimates of the rate of
lateral diffusion of phospholipids in model and biological membranes. In the liquid crystalline phase of model

bilayer membranes the lipid molecules diffuse very rapidly in the plane of the membrane [17, 18]. From the observed diffusion constant of $\simeq 3 \times 10^{-8}$ cms^2/sec, two adjacent lipid molecules will diffuse away from each other in 10^{-7} seconds. Approximately the same lateral diffusion rate has been found for the bulk of the phospholipid in biological membranes [19, 20]. However, spin labels have also provided very strong evidence for the existence of a single layer of immobilised phospholipid molecules coating the membrane proteins — the so-called 'boundary-layer' lipid [21].

All of the magnetic resonance information indicates that the motion of phospholipids in the membrane is highly anisotropic. There is a rapid axial motion of the cylindrical phospholipid molecules around the bilayer normal [22, 23, 24], but only restricted motion perpendicular to the bilayer normal. ^{2}H and spin label studies [23, 25] have shown that the amplitude of these restricted 'off-axis' motions monotonically increases from the glycerol back-bone region (which is the most rigid part of the membrane) to the interior of the hydrocarbon core (where the motion is almost isotropic). ^{13}C and ^{1}H relaxation time measurements have indicated that the rate of these restricted motions increase down the fatty acid chains from the glycerol back-bone to the centre of the hydrocarbon interior [26, 27].

Model Membrane Systems

Fig. 2 shows the structure of a typical membrane phos-phospholipid—phosphatidyl choline from yolk. The fatty acid chains vary in length and number of double bonds, but the structure shown would be typical. There are three intrinsic magnetic resonance probes in the phospholipid molecules; ^{1}H, ^{13}C and ^{31}P. The other two commonly used probes, ^{2}H and nitroxide spin labels, require specific incorporation, which makes them less amenable in biologi-

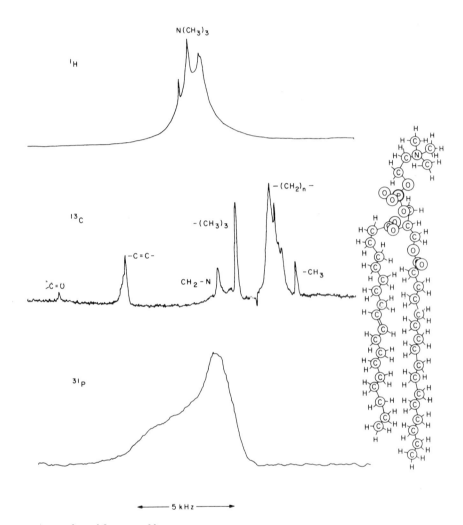

Fig. 2. ^1H, ^{13}C and ^{31}P NMR spectra of unsonicated phosphatidyl cho-line bilayers. The ^1H and ^{13}C NMR spectra were obtained on a Bruker WH-360 spectrometer on egg-yolk lecithin (100 mgs/ml, 0.1 M NaCl, T = 25°C). The ^{13}C spectra were run with 15 watts proton decoupling power. The ^{31}P NMR spectrum was obtained at 129 MHz for partially hydrated dipalmitoyl phosphatidyl choline (hydrated over pure H_2O, T = 50°C). No proton decoupling was used for the ^{31}P NMR spectrum. The spectral width is 20 kHz in all cases. The assignments of the resonances are given with respect to the structure of phosphatidyl choline given on the right. The number of carbon atoms and the number of double bonds in the fatty acid chains will vary, but the structure shown would be typical for egg-yolk phospatidyl choline.

cal membranes.

A glance at Fig. 2 shows that phospholipids contain a plethora of carbon and hydrogen atoms in different chemical and physical environments. In general it is difficult to resolve specific ^{13}C and ^{1}H signals from different regions of the membrane. The ^{1}H spectrum of the phosphatidyl choline membranes show only one resolved signal; from the choline trimethyl ammonium group.

The problem is less severe for ^{13}C spectra than for ^{1}H spectra because of the much larger ^{13}C chemical shift differences. The ^{13}C spectra of the phosphatidyl choline membranes show well-resolved resonances for the carbonyl carbons, the carbons contained in the double bond, and the methyl carbons in the fatty acid chains. Also, two carbon atoms in the polar headgroup are resolved — the tri-methyl ammonium group, and the methylene group adjacent to the nitrogen. In contrast, since the phospholipid molecules generally contain only one phosphate group, the ^{31}P spectrum is particularly simple. The spectrum is broad and asymmetric, with a pronounced low-field shoulder. This signal shape is easily understood and gives considerable information on the motion of the phosphate group in the membrane.

The frequency of the ^{31}P signal is dependent on the orientation of the external magnetic field with respect to the chemical shift tensor fixed in the phosphate group. The overall isotropic Brownian rotational rate of most biological membranes (and unsonicated model membranes) is much too slow to average this anisotropy. However, if the phosphate groups undergo a rapid but restricted motion (for instance a rotation about some axis), the anisotropic components of the chemical shift tensor will be partially averaged about this axis. The average value of the ^{31}P chemical shift anisotropy will obviously be very sensitive to the details of the restricted internal motions. 'Rapid' is defined by the reciprocal of the anisotropy of the chemical shift tensor, which is about 10^{-5} seconds.

The middle spectrum in Fig. 3 illustrates the orientation dependence of the ^{31}P NMR signals from oriented multilayers. The spectrum for the randomly oriented membranes may now be understood as simply the superposition of the spectra from the different orientations. In the superposition, the membranes perpendicular to the magnetic field will be statistically preferred, giving the apparent peak at the high-field edge of the spectrum. The average

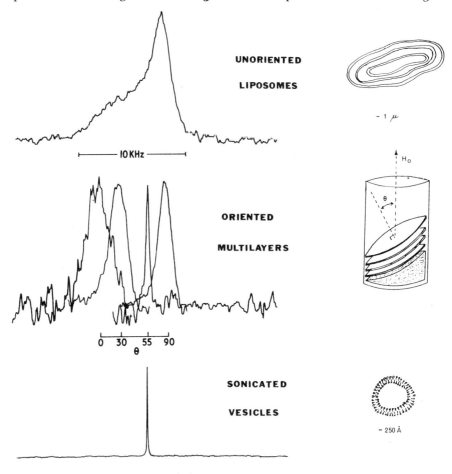

Fig. 3. ^{31}P NMR spectra from three preparations of phosphatidyl choline bilayer membranes; unsonicated aqueous dispersions, oriented multilayers and sonicated vesicles. 129 MHz.

value of the chemical shift anisotropy may be measured
from the spectra for the unoriented membranes from the
separation between the midpoints of the high and low field
shoulders, and from the spectra for the oriented multilayers as the separation between the signals from the 0°
and the 90° orientations. From the details of the orientation dependence of the position of the signals, the orientation of the axis of rotation of the phosphate group with
respect to the plane of the membrane may be determined.
The width of the signals from the oriented multilayers can
be shown to arise from the residual dipolar interaction
between the phosphorus atoms and the protons on the two
adjacent methylene groups. If the two methylene groups are
able to undergo a rapid restriction motion (for instance
a rotation about some axis) the anisotropic components of
this dipolar interaction will be partially averaged about
the axis. 'Rapid' is defined by the reciprocal of the
dipolar interaction, which is about 10^{-4} seconds.

We now have two independent methods to study the motion
of two different regions of the polar headgroup; from the
orientation dependence of the position of the ^{31}P NMR
signal we can study the motion of the phosphate group, and
from the orientation dependence of the width of the signal
we can study the motion of the adjacent methylene groups.
From the details of the orientation dependence of the
positions and the width, both the phosphate group and the
methylene groups can rapidly rotate about the normal to
the plane of the bilayer membrane in the liquid crystalline state. From ^{2}H NMR we know that, in the liquid
crystalline phase, the hydrocarbon chains are able to
rotate rapidly about the bilayer normal [23]. 'Rapid' on
the ^{2}H NMR scale is defined by the reciprocal of the quadrupole coupling constant, which is about 10^{-6} seconds.
All of this information together implies that in the
liquid crystalline state the entire phospholipid molecule
is able to rotate about the bilayer normal much faster

A.C. McLAUGHLIN ET AL.

than 10^{-5} seconds.

The question then arises, is the polar headgroup capable of independent motion about the bilayer normal, i.e. independent of the motion of the entire phospholipid molecule? Fig. 4 shows the orientation data for dimyristoyl lecithin above and below the gel-liquid crystalline phase transition, which occurs at 23.5°C.

Below the phase transition the position of the ^{31}P NMR signal is still sensitive to the orientation of the mem-

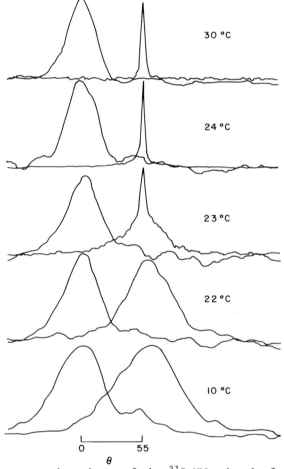

Fig. 4. Temperature dependence of the ^{31}P NMR signals from oriented phosphatidyl choline multilayers. The multilayers were hydrated over water. θ is the angle between the magnetic field and the normal to the plane of the membrane. 129 MHz.

branes; however the width of the signals has now become
rather insensitive to the orientation. This implies that,
although the phosphate group is able to rotate rapidly
about the bilayer normal, the methylene group in the glycerol back-bone is not. This then implies that the observed rotation of the phosphate group about the bilayer normal is an internal motion in the polar headgroup which is
independent of the axial rotation of the entire phospholipid molecule. Also, while the change in the orientation
dependence of the line-width of the signal occurs abruptly
at the phase transition, the average value of the chemical
shift anisotropy does not. This implies that while the
rotational motion of the entire phospholipid molecule is
severaly curtailed below the phase transition, the internal motion of the phosphate group is not substantially
affected by the phase transition.

If the model membrane preparation is sonicated, the
resulting small, single-bilayer vesicles have a rapid
enough Brownian rotation rate ($\sim 10^{-6}$ seconds) to average
out the residual anisotropies in the ^{1}H, ^{2}H and ^{31}P NMR
spectra. While this produces 'high-resolution' spectra
(i.e. see Fig. 3) which are useful for investigating the
interaction of drugs, proteins and divalent cations with
lipid bilayers, the averaging of the anisotropies makes
it difficult to exact structural information from these
spectra.

Biological Membranes

Figure 5 shows the ^{31}P NMR spectrum of erythrocyte membranes (top) and the bilayer membranes formed from the
phospholipids extracted from the erythrocyte membranes
(bottom). The average value of the ^{31}P chemical shift
anisotropy is the same for the two systems. Since at least
85 per cent of the phospholipids contribute to the erythrocyte spectrum, this implies that the protein in the
erythrocyte membrane has little effect on the motion of

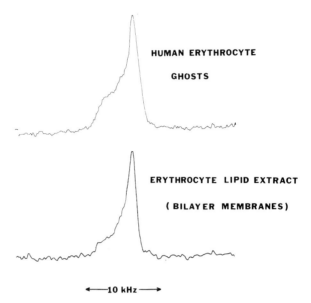

◄——10 kHz——►

Fig. 5. Top: ^{31}P NMR spectrum of human erythrocyte membranes suspended in 5 mM tris-Cl, pH 8.1, T = 7°C. Bottom: ^{31}P NMR spectrum of liposomes made from phospholipids extracted from human erythrocyte membranes. 25 mM Tris-Cl, pH 7.0, 0.2 mM EDTA, 7°C. 129 MHz. The total spectral width is 30 kHz.

most of the phosphate groups. Thus, judged by the motion of the phosphate group region, at least 85 per cent of the phospholipids in the erythrocyte membrane are in the bilayer form.

However, the spectrum for the erythrocyte membranes has a significantly higher low-field shoulder than the spectrum for the model bilayer membranes. This increase in the height of the low-field shoulder with little change in the average value of the chemical shift anisotropy is similar to the effect seen in pure phospholipids on going below the phase transition [22]. In pure lipids the effect is due to the restriction of the axial rotation of the intact phospholipid molecule. The difference between the erythrocyte membranes and the model bilayer membranes could then be explained if the protein, which does not restrict the motion of the phosphate groups, did severely

restrict the axial rotation of the intact phospholipid
molecules.

Normal and transformed cell pairs are widely used to
study the structural alterations accompanying malignancy
(e.g. [28]). The properties of transformed cells mimic
pathological deviations found in neoplastic tumour cells,
i.e. loss of contact inhibition. Since contact inhibition
is presumably mediated via a membrane interaction, it is
of interest to compare the structures of the plasma membranes of the two cells.

Fig. 6 shows the ^{1}H spectra of isolated plasma membranes from normal rat embryo cells and rat embryo cells
transformed with polyoma virus. Superimposed on the broad
background from the majority of the protons in the lipids
and the proteins, four resonances may be resolved. The

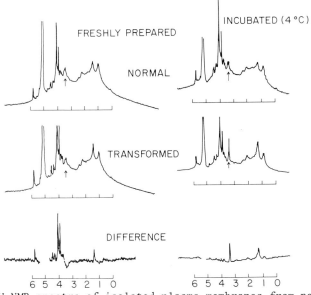

Fig. 6. ^{1}H NMR spectra of isolated plasma membranes from normal and
transformed rat embryo cells. The membranes were suspended in 0.15 N
sodium chloride, 10 mM sodium phosphate, pH 7.5, T = 10°C. The spectra on the left were taken on freshly prepared plasma membranes while
the spectra on the right were taken on the same membranes that had
been incubated for four days at 4°C. The spectra at the bottom are
the difference spectra (transformed minus normal). See text for
explanation of the assignments.

first is a peak at 3.6 ppm (down field from DDS) which is
assigned to the choline tri-methyl ammonium groups of
phosphatidyl choline and the sphingomyelin molecules. The
second and third peaks at 1.8 and 1.2 ppm are assigned
respectively to relatively 'mobile' methylene groups and
methyl groups, although it is not possible to assign them
to protein or lipid origin. The fourth is a small peak at
2.3 ppm which is assigned to the N-acetyl protons of sugar
residues at the surface of the membrane. The series of
sharp peaks from 3.7 ppm to 5.8 ppm arise from residual
sucrose in the medium. The plasma membranes are prepared
in sucrose and it is difficult to remove all traces of
sucrose without disrupting the surface of the membrane.
The peak at 5.0 ppm arises from the residual water in the
solvent.

If we compare the ^1H spectra of plasma membranes from
the normal and transformed cells there are very few differences. This is illustrated by the difference spectra
for the two systems. There is a small difference in the
amount of the choline tri-methyl ammonium signal (with the
normal cells having slightly more) and a small difference
in the amount of 'mobile' methylene groups (with the transformed cells having slightly more). Since the choline
peaks are the same width in the normal and transformed
cell membranes this small difference probably reflects a
difference in the total amount of choline headgroups
rather than a difference in the physical state of the
choline headgroups. Since the origin of the 'mobile' methylene peak is obscure, the interpretation of the small differences in this peak is not clear. If the membranes are
incubated for four days at 4°C, we see a large difference
between the normal and the transformed membranes. The ^1H
spectrum for the normal plasma cell membranes has not significantly altered during the incubation. However, the
spectrum for the transformed plasma membranes is significantly different.

These changes are illustrated in the difference spectrum. Clearly the transformed cell membranes have a large
'mobile' (i.e. very narrow) choline tri-methyl ammonium
peak, which is absent in the normal plasma membranes. Also
the transformed plasma membranes have more of the 'mobile'
methylene signal. Qualitatively the same effects of incubation at a low temperature are seen by raising the temperature of the sample (to 30°) for a short period of
time (about one hour).

The ^{31}P NMR spectra of isolated plasma membrane from
normal hamster NIL 8 cells and hamster cells transformed
with hamster sarcoma virus are shown in Fig. 7. The spectra of the normal and the transformed plasma membranes are
identical. Also, spin label studies have shown little, if
any, differences in the fluidity of the hydrocarbon region

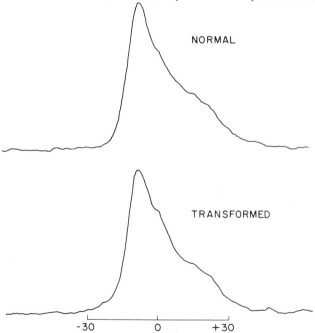

Fig. 7. ^{31}P NMR spectra of isolated plasma membranes from normal
hamster NIL 8 cells and hamster cells transformed with hamster sarcoma virus. The membranes were suspended in 0.15 N sodium chloride,
20 mM sodium borate, pH 7.5. The scale is in parts per million from
phosphoric acid. Frequency is 129 MHz. T = 10°C.

of normal and transformed cell pairs [29].

Chapman [30], Inesi [31] and Chan [32], have found that
the absence of a narrow choline tri-methyl ammonium peak
was a good indication of the integrity of erythrocyte and
sarcoplasmic reticulum membranes. Treatments which disrupted the integrity of the membranes, such as limited
trypsin digest, removal of spectrin from the erythrocytes,
sonication or exposure to high temperatures produce a
narrow choline tri-methyl ammonium peak. The interpretation was that, in the intact membranes, the choline group
was immobilized by an interaction with neighbouring molecules, presumably proteins. Treatments which disrupted
this interaction then generated a mobile choline group.
The above data suggests that the interaction between the
lipid headgroups and the membrane proteins is more labile
in the transformed plasma cell membranes than in the normal cell membranes. However the phosphate group region and
the hydrocarbon region [29] of the normal and transformed
cell plasma membranes appear to be very similar. Thus, the
transformation process appears to be a subtle effect on
the stability of the lipid–protein interaction rather than
a gross alteration in the structure of the membrane.

Conclusions

Magnetic resonance techniques are well-suited to extract dynamic information about model and biological membranes. They have shown that the lipid bilayers are fluid
yet highly anisotropic systems, and that the majority of
the lipids in biological membranes have motional properties similar to the model lipid bilayers. Also, magnetic
resonance techniques may be used to investigate lipid–
protein interactions in biological membranes.

Perhaps the most interesting region of the membrane is
the protein and the 'boundary-layer' lipids. However since
the motion of the protein in the membrane may be severely
restricted, the NMR signals from these components are

rather broad. The application of the sophisticated pulse sequences developed by Waugh and coworkers [33] for averaging residual solid state interactions will hopefully provide more information on these more rigid, but functionally more interesting, regions of the membrane.

References

1. Engelman, D.M. (1970). *J. Mol. Biol.* **47**, 115.
2. Steim, J.M., Tourtellotte, M.E., Reinert, J.C., McElhaney, R.M. and Rader, R.T. (1969). *Proc. Nat. Acad. Sci. (U.S.A.)* **63**, 104.
3. Stoeckenius, W. and Engelman, D.M. (1969). *J. Cell. Biol.* **42**, 613.
4. Singer, S.J. and Nicolson, G.E. (1972). *Science* **175**, 720.
5. Chapman, D., Williams, R.M. and Ladbrooke, B.D. (1967). *Chem. Phys. Lipids* **1**, 445.
6. Trauble, H. and Overath, P. (1973). *Biochim. Biophys. Acta* **307**, 491.
7. Berden, J.A., Cullis, P.R., Hoult, D.I., McLaughlin, A.C., Radda, G.K. and Richards, R.E. (1974). *FEBS Lett.* **46**, 55.
8. Penkett, S.A., Flook, A.G. and Chapman, D. (1968). *Chem. Phys., Lipids* **2**, 273.
9. Warren, G.B., Toon, P.A., Birdsall, N.J.M., Lee, A.G. and Metcalfe, J.C. (1974). *Proc. Nat. Acad. Sci. (U.S.A.)* **71**, 622.
10. Overath, P. and Trauble, H. (1973). *Biochem.* **12**, 2625.
11. Linden, C.D., Wright, K.L., McConnell, H.M. and Fox, C.F. (1973). *Proc. Nat. Acad. Sci. (U.S.A.)* **70**, 2271.
12. Bretcher, M. (1973). *Science* **181**, 622.
13. Wilkins, M.H.F., Blaurock, A.E. and Engelman, D.M. (1971). *Nature, New Biology* **230**, 72.
14. Engelman, D.M. (1971). *J. Mol. Biol.* **58**, 153.
15. Henderson, R. (1975). *J. Mol. Biol.* **93**, 123.
16. Zaccai, G., Blasie, J.K. and Schoenborn, B. (1975). *Proc. Nat. Acad. Sci. (U.S.A.)* **72**, 376.
17. Lee, A.G., Birdsall, N.J.M. and Metcalfe, J.C. (1973). *Biochem.* **12**, 1650.
18. Devaux, P. and McConnell, H.M. (1972). *Am. Chem. Soc.* **94**, 4475.
19. Scandella, C., Devaux, P. and McConnell, H.M. (1972). *Proc. Nat. Acad. Sci. (U.S.A.)* **69**, 1056.
20. Sackman, E., Trauble, H., Galla, H.J. and Overath, P. (1973). *Biochem.* **12**, 5360.
21. Jost, P.C., Griffith, O.H., Capaldi, R.A. and Vanderkooi, G. (1973). *Proc. Nat. Acad. Sci. (U.S.A.)* **70**, 480.
22. McLaughlin, A.C., Cullis, P.R., Hemminga, M., Hoult, D.I., Radda, G.K., Ritchie, G.A., Seeley, P.J. and Richards, R.E. (1975). *FEBS Letts.* **57**, 213.
23. Seelig, A. and Seelig, J. (1974). *Biochem.* **13**, 4839.
24. Jost, P., Libertini, L.J., Herbert, V.C. and Griffith, O.H. (1971). *J. Mol. Biol.* **59**, 77.
25. Hubble, W. and McConnell, H.M. (1971). *J. Am. Chem. Soc.* **93**, 314.
26. Levine, Y.K., Birdsall, H.J.M., Lee, A.G. and Metcalfe, J.C. (1972). *Biochem.* **11**, 1416.

27. McLaughlin, A.C., Podo, F., and Blasie, J.K. (1973). *Biochem.,
    Biophys. Acta* **458**, 1.
28. Ponten, J. (1976). *Biochem., Biophys. Acta* **458**, 397.
29. Gaffney, B.J. (1975). *Proc. Nat. Acad. Sci. (U.S.A.)* **72**, 669.
30. Chapman, D., Kamat, V.B., de Gier, J. and Pennkest, S.A. (1968).
    J. Mol. Biol. **31**, 101–14.
31. Davis, D.G. and Inisi, G. (1972). *Biochem., Biophys. Acta* **282**,
    180.
32. Sheetz, M.P. and Chan, S.I. (1972). *Biochem.* **11**, 548.
33. Haeberlen, V. and Waugh, J.S. (1969). *Phys. Rev.* **185**, 420.

PHOSPHORUS NMR IN LIVING TISSUE

P.J. SEELEY, P.A. SEHR, D.G. GADIAN,
P.B. GARLICK and G.K. RADDA

Department of Biochemistry, University of Oxford

Introduction

The understanding of the control of cellular bioener-
getics requires the precise definition of many factors
that modulate the activities of different biochemical pathways that are involved in the production and utilization
of 'energy' in living systems. Very crudely, we can view
energy production and demands by a cell as being centred
around the availability of ATP. In Scheme I we illustrate
some of the major energy producing and utilizing pathways.
Energy input is largely derived from either oxidative metabolism of substrates (possibly through the intervention of
H^+ ion gradients) or from the metabolism of sugars through
glycolysis. Supply for this pathway is often derived from
the 'glycogen store'. The oxidative and glycolytic paths
are strictly linked through a series of specific control
mechanisms, as are the processes that impose a demand on
the energy supply. We are faced with an interesting para-
dox if we are to define biological control at the molecular level. On the one hand intermolecular interactions
that are important in regulation are best studied on isolated, well-characterized components. On the other hand,
control of a living system relies entirely on the integration of these components. It was with this in mind that
some years ago we examined ways of translating observations

SCHEME I

Bioenergetic Processes

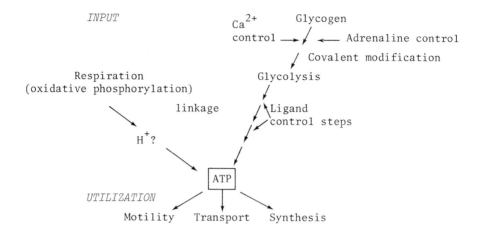

made *in vitro* to *in vivo* situations [1].

   Since phosphorus-containing metabolites were readily observed by ^{31}P NMR [2] we considered the possibility that this technique could be used to observe molecular events in intact biological preparations. This was made possible by advances in instrumentation [3] and a detailed programme of collaborative research. We have demonstrated some of the potentials of ^{31}P NMR by studying metabolite turnover in isolated glycogen particles [4,5,6] and more importantly in whole muscle [1,4,6].

   In this presentation we examine some of the contributions ^{31}P NMR has made in our biochemical studies and also try to give examples of how we envisage using ^{31}P NMR in biological and medical research.

   It is convenient, for our present discussion, to break down the various biological control mechanisms into 'control elements' (Table 1) that are particularly amenable for study by ^{31}P NMR in intact biological tissue. To this we should add that we have deliberately chosen to examine specialised tissues and organs, where differentiation has

TABLE 1

'Control elements' in biology

---

1. Ligand regulation of key enzymes
2. Covalent modification of key enzymes
3. Hormonal control
4. Cations
5. H^+
6. Availability of substrates
   a. concentrations (fluxes)
   b. compartments

---

emphasized particular biological functions, but other systems, like bacteria, can also be studied in this way [7].

Compartmentation of Metabolites in Skeletal Muscle

After the initial demonstration that ^{31}P NMR signals
can be observed from intact, non-perfused skeletal muscle
[4], several lines of investigation became feasible. The
energetic requirements of contracting muscle have been examined and this work is summarised in detail in this volume [8]. Different muscles have been compared by us [9–11]
and by the Chicago group (for a summary including references to earlier work see ref. [12]) who have also studied
diseased muscle [12].

Here we briefly examine the problem of metabolite compartmentation (cf. Table 1) partly to introduce the ^{31}P
NMR method and partly to emphasize one of the new types of
information that this method provides. Typical ^{31}P NMR
spectra of non-perfused rat Vastus lateralis muscle are
shown in Fig. 1. The resonances corresponding to the three
phosphate groups of ATP, phosphocreatine, inorganic phosphate and sugar phosphate can be readily identified [4].

Apart from giving the quantities of the individual metabolites, which agree with conventional analytical data
[4,8,9,10,11,13] the spectra in Fig. 1 contain valuable

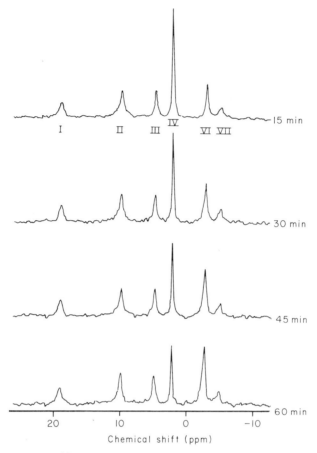

Fig. 1. Series of ^{31}P NMR spectra of a rat Vastus lateralis, recorded as a function of time after excision. The muscle was in Ringer solution at 15°C throughout the course of the experiment. Each spectrum represents the accumulation of 400 scans repeated at 2s intervals, and the times given are the midpoints of each accumulation. Peaks I, II and III arise from the β, α, and γ phosphorus atoms respectively of ATP, IV from phosphocreatine, VI from inorganic phosphate and VII from sugar phosphates.

new information. The first is that the position of the inorganic phosphate resonance faithfully reflects the intracellular pH [4,7,10,14]. Chemical shifts might result from factors other than the ionization state of the phosphate group, e.g. interaction with metal ions or binding to proteins. We have rigorously excluded the former as being

important for inorganic phosphate and have no evidence
that any significant frequency effects result from binding. Even if they do, one is often more interested in
changes in pH such as can be observed when the energy pool
of the muscle gradually runs down under anoxic conditions
(Fig. 1 and ref. 4). The second feature of the spectra in
Fig. 1 is that from the frequencies of the three ATP signals and assuming that ATP experiences the same pH environment as inorganic phosphate, we can conclude that almost
all of this nucleotide is complexed with a divalent cation,
most likely to be Mg^{2+} [4,8]. In our view, the information
we can derive about the cellular environment of the observed components is one of the most important features of
the NMR method.

In this respect a multi-component inorganic phosphate
signal is frequently observed during metabolic run-down of
muscles in anoxia [4,6,9—11]. This suggests that muscular
inorganic phosphate experiences several pH environments
under these conditions [11,16]. The exchange of ortho-
phosphate between its various environments must have a
rate of $\leq$ 50 Hz, and it seems probable that the environments are separated from each other by membrane barriers
[16].

These observations led us to examine the effects of
perturbing the intracellular pH using an acetate buffer
[16]. Fig. 2 shows the spectra of Vastus lateralis muscle
that had been bathed in acetate buffer, pH 5.2, for twenty
minutes before being placed in the spectrometer. The interesting feature of these spectra is the gradually increasing
amplitude of a 'new' signal of chemical shift - 1.05 ppm.
From this we have concluded [16] that inorganic phosphate
in the muscle is present in two types of solution, one at
the usual muscle pH ($\simeq$ 7.0) and another at about pH 5.8.
The time course of the appearance of the 'low pH' pools
suggests that its magnitude is not dependent upon the extent to which acetate has penetrated macroscopically

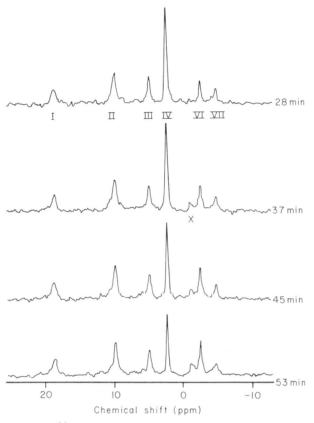

Fig. 2. Series of ^{31}P NMR spectra of a rat Vastus lateralis, recorded as a function of time after excision. The muscle was bathed in acetate buffer pH 5.2 for 20 min, and then placed in this buffer into the spectrometer. The temperature was 15°C throughout. Each spectrum represents the accumulation of 200 scans repeated at 2s intervals, and the times given are the midpoints of each accumulation. Peak assignments are as for Fig. 1, and the additional peak X is assigned to inorganic phosphate at low pH.

through the muscle volume; a steady state is reached when approximately 35 per cent of inorganic phosphate is in the 'low pH' compartment and 65 per cent in the 'high pH' compartment. This in turn suggests that inorganic phosphate in one type of environment in the muscle is accessible to acetate, whereas that in another type is not [16]. We have observed similar behaviour for sugar-phosphate (but not ATP or phosphocreatine) and interpreted it as demonstrating compartmentation of inorganic phosphate and sugar-phosphate [16]. How such compartmentation may account for the observed activity of phosphorylase b in resting muscle has been discussed elsewhere [1]. It is also thought that glycolytic activity may be associated with the sarcoplasmic reticulum [17,18] and our suggestion regarding compartmentation of inorganic phosphate and glucose 6-phosphate [16] is compatible with this view.

H$^+$ Ion Gradients and the Storage and Release of Adrenaline

We now turn to an important biochemical problem in relation to the control of cellular metabolism, namely that of the storage and release of adrenaline. In the adrenal medulla, adrenaline is stored in granules (storage vesicles) within the so-called chromaffin cells. This storage system is analogous to that in sympathetic nerves. The isolated storage vesicles (chromaffin granules) are characterised by their high content of adrenaline (0.55 molar) and ATP (about 0.13 molar) together with an acidic protein [19]. The fundamental problems in understanding this system relate to the nature of the internal storage complex, the mechanism of adrenaline release and the way in which adrenaline is accumulated within the granules to achieve these high concentrations [20]. The problems are similar in nerve endings and the storage and release mechanisms in the adrenal medulla are often regarded as good models for the adrenergic nerve. The membranes of the chromaffin granule are characterised by several enzymic activities that include an ATP-ase, an NADH-oxidase, cytochrome and a dopamine-β-hydroxylase [19,20,21,22]. On the basis of several biochemical experiments that included the use of uncouplers [21,23,24,26], measurements of the distribution of a neutral amine (methylamine) [25] across the granule membrane and fluorescence studies using a 'probe', 1-anilino-naphthalene-8-sulphonate [21,23], we have concluded that the ATP-ase is most likely to be an electrogenic pro-

ton pump [20,25,26]. We have also noted that the pH inside
the chromaffin granules was much lower than that of the
cytoplasm [20,21,23,25]. Catecholamine uptake into these
vesicles is driven by the proton translocating ATP-ase
[20,24,25]. This conclusion was based on the observation
that catecholamine transport is also inhibited by mitochondrial uncouplers [21,24]. Because of the high concentration of ATP within these granules, this system offers
the unique opportunity to use ^{31}P NMR in our biochemical
studies. Fig. 3 shows ^{31}P NMR spectra obtained when whole
adrenal glands were placed in the spectrometer together

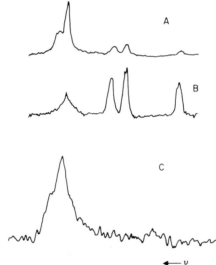

Fig. 3. ^{31}P NMR (at 36.4 MHz) spectra of chromaffin granules (A) of
adrenal glands (B) and of adrenal glands after perfusion with Ca^{2+}
and X537A (C) (Sweep width 1.25 kHz).

with spectra recorded from suspensions of isolated chromaffin granules. From these spectra we can conclude the following: first, the ATP resonances in the whole adrenal
gland are similar to those in the isolated granules indicating that no damage has occurred to the storage complex
during the isolation of the vesicles [28]. Second, when a
whole gland is perfused with either acetylcholine or Ca^{2+}
and an ionophore to release adrenaline, the ATP is also lost

(and is washed out by the perfusion medium). This is consistent with the proposed exocytotic mechanism for the
release of adrenaline [29]. Third, the resonances of the
α, β and γ phosphates of ATP in these spectra give an
unusual pattern of chemical shifts. Two factors contribute
to this. (i) ATP interacts with the positively charged
catecholamines inside the granules and (ii) the intragranular pH is close to 5.5.

It is important to note that in this kind of situation
the frequency of the γ-phosphate resonance can only be
used to measure the internal pH after careful calibrations
that take into account all the possible interactions within
the complex [28]. This can be done in several ways. We may
use the known chemical composition of the internal complex
and observe the effect of the different components on the
ATP resonances in a solution at identical concentrations
to those within the granule. We have carried out such calibrations as a function of pH [28]. In these we attempted
to account for the presence not only of adrenaline, ATP
and the acidic protein but all the other cations like Mg^{2+},
Ca^{2+} and K^+, that are present in the granules [19]. This
approach is only valid if the analytical data on the composition of the internal granule content are correct and
accurate and if there are no variations between different
adrenal glands. There is some uncertainty about both of
these [19]. Our NMR experiments on a large number of glands
and isolated chromaffin granules do, however, show a remarkable consistency. The only variations we have seen in
the spectra were when we compared glands from different
animals [28] and even then only the widths and not positions of the ATP phosphate resonances showed differences.

To exclude the analytical uncertainties, in the alternative method we subjected chromaffin granules to hypo-
osmotic lysis, isolated the granule matrix and reconcen-
trated the obtained material to the same concentration as
in the intact system. We then measured the pH of this solu-

tion with a platinum electrode (since amines interfere
with glass electrodes) and observed the ^{31}P NMR spectra as
a function of pH. Both this and the first method of calibration showed that the intragranular pH is 5.6 ± 0.1.

We are now ready to examine the suggestion, based on
biochemical experiments, that the ATP-ase is an inwardly-
directed proton pump. Fig. 4 shows an experiment in which
we have isolated granules, observed the ^{31}P NMR spectrum
and added ATP and Mg^{2+} on the outside. The resonances
derived from the ATP inside the granules and those from
ATP outside can be clearly distinguished because of the
different environmental conditions. The ATP outside is
being hydrolysed, as is demonstrated by the gradual de-

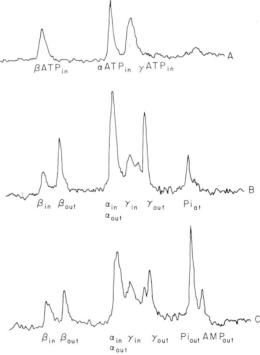

Fig. 4. ^{31}P NMR spectra of chromaffin granule samples. Chromaffin
granules in 0.8 ml of 120 mM KCl-40 mM Mes-185 mM sucrose (pH 6.44)
were incubated at 25°C. (A) Chromaffin granule suspension (0.8 ml).
(B) Chromaffin granule suspension 1-5 min after the addition of 0.1 ml
of 100 mM ATP-100 mM M_gSO_4 (pH 7). (C) Chromaffin granule suspension
34-38 min after the addition of Mg-ATP.

crease of the 'outside' ATP resonances and the concomitant
increase in the inorganic phosphate signal. At the same
time there is no change in the magnitude of the 'inside
ATP' signals, but if we look carefully at the γ-ATP peak
from 'ATP inside' (Fig. 5) we see that its position shifts

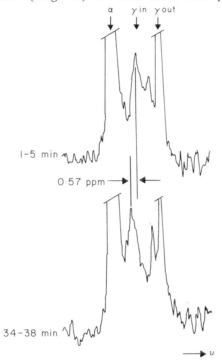

Fig. 5. ATP-dependent shift in the Y peak of intragranular ATP. The
spectra shown in Figures 4B and 4C are reproduced to show the shift in
the position of the γ-phosphate peak of ATP contained within the chromaffin granules.

in a manner that indicates that the pH inside the granules
has decreased by about half a pH unit [25]. The drop in
the intragranular pH is only observed in the presence of
permeable anions, like chloride, but not when the anion is
sulphate or when there are no anions present (i.e. in a
sucrose medium) (Fig. 6). The time course of the change
can be easily followed (Fig. 6). Since the buffering capa-
city of the granule matrix is very high, this half unit
change in internal pH represents a substantial proton

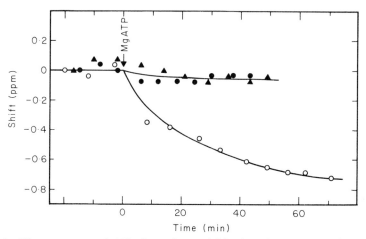

Fig. 6. Time course of ATP-dependent shift in the γ peak of intragranular ATP. For each series, 0.2 ml of chromaffin granules in 0.3 M sucrose-10 mM Hepes (pH 7.0) were diluted into 0.6 ml of 160 mM KCl-53 mM Mes-150 mM sucrose (pH 6.34) and incubated at 25°C. At t = 0 0.1 ml of 100 mM ATP-100 mM MgSO₄ (pH 7) was added, and spectra were taken periodically. The final solutions were 120 mM KCl-185 mM sucrose-40 mM Mes-2.5 mM Hepes (pH 6.37) (O), 80 mM K₂SO₄-185 mM sucrose-30 mM Mes-2.5 mM Hepes (pH 6.41) (O), and 375 mM sucrose-30 mM Mes-10 mM Hepes (pH 6.38) (Δ). The position of the γ-phosphate peak is plotted vs. the time at which the accumulation of the spectrum was completed.

influx. The medium requirements for the observed pH change clearly indicate thatthe proton translocation is electro-genic [25]. Thus movement of the charged hydrogen ions generates an opposing electrical potential that limits the amount of proton transport that can occur. This electrical restraint can be relaxed by adding a permeant anion to the medium and permitting a coupled proton-anion influx. Since H^+ anion uptake raises the internal osmolarity and will cause lysis unless the granules are in a hypertonic medium [26], the observation of extensive proton uptake requires that the external medium both contain a permeant anion and be hypertonic. Our NMR observations have been confirmed by two independent methods [25,26]. Finally we note that active uptake of catecholamines can take place in both chloride-containing or sucrose media. This implies that both the pH difference and the trans-membrane differences in electrical potential may drive catecholamine uptake.

Studies on Perfused Heart

In the last two sections we have demonstrated that ^{31}P NMR can give specific new information about metabolic events and their mechanisms in intact tissue. How we define an 'intact tissue' depends on the particular problems one wishes to study. For example, from the physiological point of view our experiments on skeletal muscle can be regarded as being done on 'dying muscle' [8], yet of course the bio-chemical events associated with hypoxia are of interest in relation to the control of oxidative versus non-oxidative energy production. The ideal situation is to be able to ex-periment on a 'functioning' system and for this reason isolated organs offer many possibilities.

The heart can be kept in a functional steady state, if perfused correctly, for many hours and we have therefore chosen to extend our study towards cardiac metabolism.

In our initial experiments, we demonstrated that NMR spectra can be observed from a small rat heart which was rapidly cooled after removal [30]. At 4°C the heart stops beating and energy utilisation and oxygen demand are expec-ted to be minimal as a result of decreased metabolic rate and inhibition of mitochondrial adenine nucleotide trans-locase, thus the tissue should have established a resting metabolic state. Fig. 7 shows the spectrum obtained from such a preparation. The peaks can be assigned as before [4,30] and the values of the integrals (given in the legend to Fig. 7) can be measured after about half an hour accu-mulation. A notable feature of these integrals is that the three signals corresponding to ATP are in the ratio $\beta:\alpha:\gamma$ = 0.5: 1.5: 0.9. This implies that the γ peak contains a contribution from ADP and that the α peak in addition to ADP also contains resonances from other pyrophosphates, possibly NAD and related molecules. The frequency of the inorganic phosphate resonance shows that the intracellular

P.J. SEELEY ET AL.

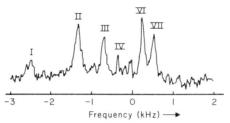

Fig. 7. ^{31}P NMR spectrum of a 170 mg rat hearts recorded at 129 MHz without proton irradiation. Temperature 4°C. The pulse interval was 2s and the pulse angle 60°. Sweep width 5 kHz. This spectrum represents a sum of spectra from two separate hearts (total no. of scans 2240). Such summary was justified by the similarity of the individual spectra obtained from each of the hearts (see [30]) and was done to obtain accurate integrations. The relative integrals for the peaks were I:1; II:3.0; III:1.8; VI:2.4; VII:1.6.

pH of the resting heart muscle is ~ 7. After the temperature of the NMR sample cavity was raised to 30°C and about 10 minutes was allowed for equilibration and for the stoppage of intermittent contractions of the heart, spectra were recorded by accumulating sets of 200 scans successively (Fig. 8). These spectra show a steady increase in the inorganic phosphate level and the run-down of the energy store of the cardiac tissue. The significant result is that the inorganic phosphate resonance progressively shifts to lower frequency, indicating that the intracellular acidity has dropped to about pH 6 in approximately 15 minutes.

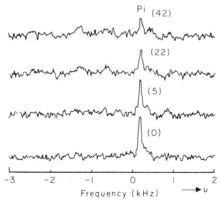

Fig. 8. ^{31}P NMR spectra of a 200 mg rat heart at various times after warming to 30°C. Numbers in parentheses are frequencies of the maximum ordinate of inorganic phosphate signal in Hz.

We have argued [30] that since the inorganic phosphate
signals from the hypoxic ischaemic states are significantly
shifted from those observed in normoxic tissue, ^{31}P NMR
provides a diagnostic tool for abnormal metabolic conditions associated with cardiac infarct. Tissue acidosis is
clearly a measure of increased glycogenolysis and the con-
sequential lactate production that may well be responsible
for tissue damage in cardiac ischaemia. To evaluate the
validity of such possibilities it is, of course, necessary
to have a direct comparison between a normal functioning
heart under physiological conditions and the damaged system.

It was therefore necessary to perfect and extend the
perfusion technique so that a small rat heart (the size
being dictated by the 8 mm internal diameter of the sample
tube we use in the superconducting magnet) could be kept
in a functional steady state in the spectrometer.

The apparatus designed and built for heart perfusion is
given in Fig. 9. Special care was taken to thermostat all
the containers and tubes so that the heart could be kept
at 37°C. The perfusion medium, the nature of the flexible
connections (so that, for example, no oxygen loss from the
buffer occurs) and the conditions used in removing the
heart are all important technical details that determine
the success of such experiments. Their detailed discussion,
however, is inappropriate for this presentation.

It is sufficient to note that in our initial experiments
where we were only able to keep the heart beating for about
an hour or so, the levels of the various metabolites observed by NMR did not stay constant throughout the experiments (Fig. 10). Initially we noted the build up of phosphocreatine (Fig. 10 (i)-(iii)) but after about 30 minutes,
inorganic phosphate started to build up again and the creatine phosphate was slowly being used up (Fig. 10 (iv-v)).
Having overcome all the technical problems of perfusion we
have demonstrated that when the heart was kept functioning

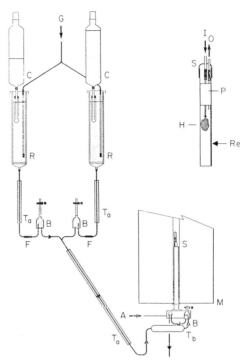

Fig. 9. Heart perfusion apparatus used for NMR measurements. Glassware is maintained at constant temperature by rapid flow of thermostatted water. The right hand reservoir and bubble trap are auxiliaries used for experiments involving potassium arrest, substrate alteration, etc. *Key:* G - Gas Input (95% O_2, 5% CO_2); C - Constant head device; R - Reservoir; T_a - Thermostatted glass tube; T_b - Thermostatted glass tube, helical; F - Flow control; B - Bubble trap; A - Input for thermostatting air; M - Superconducting magnet; S - Sample tube; I - Aortic input for perfusion fluid; O - Output for coronary flow; P - Teflon support post for cannula; H - Heart; Re - Fine reference capillary; S - Rubber sealing cap.

in a steady state for 5—6 hours very reproducible spectra could be observed [31] and a typical spectrum is shown in Fig. 11. The intracellular pH in the perfused heart is 7.4 and the ratio of creatine phosphate to ATP is close to 2. The concentration ratios for perfused heart are in broad agreement with those measured for perchloric acid extracts (Table 2). The high value of creatine phosphate to ATP and low inorganic phosphate to ATP ratios are indices of efficient cardiac respiration. These values remained constant

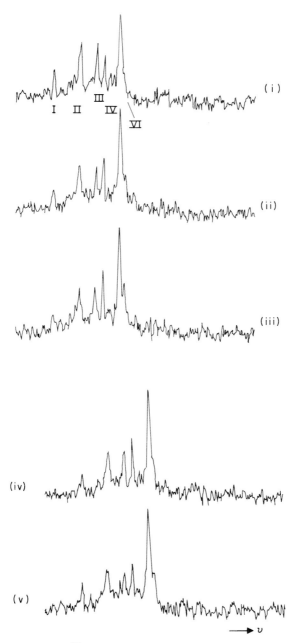

Fig. 10. Consecutive ^{31}P NMR spectra of a 150 mg rat heart at 37°C. Each spectrum is the sum of 200 sweeps taken at 2s intervals. Sweep width 10 kHz. The heart was perfused with filtered Krebs-Henseleit solution gassed at 37°C with 95% O_2. For peak assignments see Fig. 1.

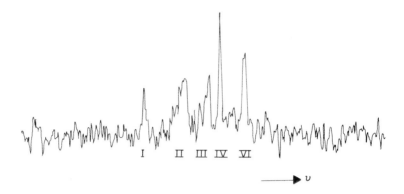

Fig. 11. 129 MHz phosphorus NMR spectrum of a 160 mg rat heart perfused by the Langendorff technique at 37°C. Sweep width 10 kHz. 1000 60° radio-frequency pulses were applied at 4s intervals. Peak assignments given in Fig. 1.

throughout the perfusion period and are therefore, along with heart and coronary flow rates, a further indication of a functional steady state of the heart. It is particularly encouraging that our measurements on correctly perfused hearts (using both direct NMR recording and chemical analysis after extraction) are highly reproducible and indeed that ^{31}P NMR provides a very sensitive measure of how 'healthy' our heart preparation is.

The observations summarised in Fig. 12 follow from this. Here we show phosphorus spectra of hearts subjected to periods of global ischaemia with and without return of flow. The levels of sugar phosphate and inorganic phosphate are raised during ischaemia [31] as a result of activation of glycogenolysis, but after 15 minutes of ischaemia reperfusion results in the total recovery of the original spectrum. The one observable difference is that the sugar phosphate level remains higher after the ischaemic period than in the initial perfusion, although the other ratios of creatine phosphate, inorganic phosphate and ATP recover their original value. There is no doubt, therefore, that phosphorus NMR provides a new way of looking at functioning and diseased cardiac tissue. Similar observations have been

TABLE 2

Integrals of phosphorus signals relative to β-ATP

160 mg rat hearts direct observation	β-ATP	α-ATP	γ-ATP	Creatine phosphate	Inorganic phosphate	Sugar phosphate and nucleotide monophosphate
4s pulse intervals	1	3.6±0.5	1.7±0.3	1.9±0.3	1.8±0.3*	n.o.
2s pulse intervals	1	1.6±0.3	1.1±0.2	1.4±0.3	1.6±0.3*	n.o.
800 mg rat hearts 3% perchloric acid extraction						
4s pulse intervals	1	1.5±0.2	1.2±0.2	1.7±0.3	2.2±0.3	0.9±0.2

n.o. = not observable

*Corrected for inorganic phosphate in Krebs-Henseleit buffer.

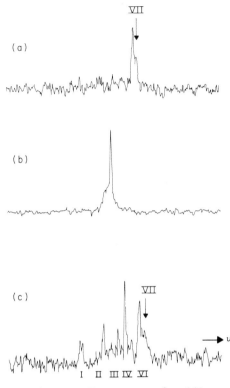

Fig. 12. 129 MHz phosphorus NMR spectra of a 160 mg rat heart under various ischaemic conditions. (a) 20 minutes ischaemia without reflow (600 60° pulses at 2s intervals. 10 kHz sweep); (b) Wide sweep spectrum collected after spectrum (a) (6000 30° pulses at 0.5s intervals. 50 kHz sweep); (c) 15 minutes ischaemia plus reflow (1200 60° pulses at 2s intervals. 10 kHz sweep). Assignments: Fig. 1.

made and reported by the Baltimore Group [32] who have also exhibited a poster at this meeting. It is gratifying to see general agreement between the two sets of observations. The differences seen are likely to be the result of the different experimental conditions employed. For example, the lack of inorganic phosphate signal reported by Hollis *et al*. [32] may well arise because this group did not include inorganic phosphate in their perfusion medium.

In our experience function is more rapidly impaired in the absence of added phosphate. In many ways the demonstration of such differences helps our understanding of the

factors that are important for the physiological function
of cardiac tissue and gives us more confidence that ^{31}P
NMR has a diagnostic value in cardiac disorders.

A Model Kidney Transplant

It is obvious that the kinds of experiments we have described here could be extended to study almost any organ
that can be perfused in an isolated state. We were therefore particularly pleased when as a result of our work on
whole tissue NMR we were approached by the surgical team
of Robert Sells and Peter Bore at the Renal Transplant Unit
of the Liverpool Royal Infirmary. They have been concerned
in establishing some relation between the metabolic state
of an isolated human kidney and the viability of the tissue
for transplant operations. In their animal experiments the
physiological function of the ischaemic kidney is conven-
tionally measured by inducing injury in one kidney, by
renal artery clamping, followed by contra-lateral nephrectomy, and observing the survival of the experimental animal. Chemical and biochemical analysis on the one kidney
removed can then be used to try and assess the relation
between organ viability after ischaemia and the metabolic
state of the tissue. The possibility that ^{31}P NMR could be
used to examine the kidney without damaging it provides a
new dimension in renal research. We have therefore decided
that in collaboration with the Liverpool surgeons we should
attempt an experiment that is as close to a kidney transplant operation as is possible. Let us follow through what
happens in such an operation in our animal model experiment
[33]. After death of the donor is established the kidney
is cooled during the operation, taken out and placed on
ice as quickly as possible. Two periods appear to be critical for graft viability, the time that the kidney spends at
body temperature without circulation before its removal
(the warm ischaemic period) and the time of the cold period
on ice, which on the whole cannot be longer than 24 hours.

In our animal experiments the warm ischaemic period was extremely short, less than a minute. After that the kidney was chilled to 4°, the ^{31}P NMR spectrum was recorded within 4 minutes of nephrectomy. The spectrum corresponding to this state still shows a great deal of ATP, a small amount of creatine phosphate and some other components as is given in Fig. 13. We note that a metabolite that corresponds to a phosphodiester (that has been observed in some muscle preparations [8-12]), is very abundant in the kidney. The level of AMP is also much higher in renal tissue than in

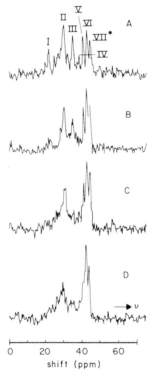

Fig. 13. ^{31}P NMR spectra of a non-perfused rat kidney at 4°C. The kidney was excised, cooled for ten seconds in ice cold Krebs buffer and then transferred to the thermostatted NMR tube. Spectra were recorded at 129 MHz in the Fourier Transform mode by applying 200 70° radio frequency pulses at 2s intervals. No field frequency lock was employed. Spectral accumulation was started 1 min (A), 15 min (B), 41 min (C), and 57 min (D) after the onset of ischaemia. Peak assignments as Fig. 1. V is a signal from unidentified phosphate. VII* is likely to be mostly AMP and some sugar phosphates.

the other tissues we have examined. The subsequent spectra
in Fig. 13 show how, even at 4°C, the energy pool of renal
tissue is gradually depleted. The time scale for this decrease is comparable to the time normally required for
transporting the isolated non-perfused kidney from the
donor to the operating theatre of the recipient. The relatively rapid utilization of ATP under anoxic/ischaemic conditions is not particularly surprising as in the kidney
the initial creatine phosphate level is rather low. A significant difference between renal tissue and muscle is
that apparently even during two hours of cold ischaemia no
changes in cellular pH are detected in the former by the
position of the inorganic phosphate resonance. This is
entirely consistent with the relatively low glycogen content and the slow rates of glycolysis in renal tissue at
4°C.

We now place the ischaemic kidney in the spectrometer
and essentailly link it in to the circulatory system of a
live 'assist' rat. The kidney in the spectrometer is perfused by the blood circulation of the anaesthetised animal
as is sketched out in Fig. 14. In this method an arterial
cannula from the anaesthetised adult (400 g) rat is linked
via a peristaltic pump and a 37°C water jacket to the NMR
tube where it joins the arterial cortex cannula (external
diameter 0.75 mm) of the perfused kidney removed from a
small (50 g) rat. The venous effluent drains into the NMR
tube and blood is sucked out by a venous cannula and pump.
The venous line runs from the NMR tube (again kept at 37°C)
via a second pump and bubble trap back to the vena cava of
the assist rat. Fig. 15 shows ^{31}P NMR spectra of the small
kidney after commencing perfusion. It can be seen that ATP
is re-synthesized after about one hour of perfusion and
its level remains relatively stable for several hours.
After 133 minutes perfusion was stopped for a period of 12
minutes, the kidney being kept at 37°C in the spectrometer.
The ischaemic spectrum was then recorded (Fig. 15c). Re-

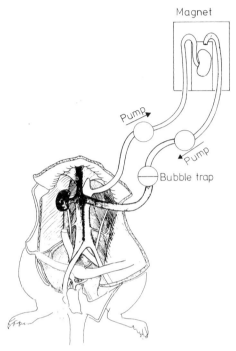

Fig. 14. Perfusion of a small rat kidney using the blood circulation of a large ('assist') rat. The right kidney of the large rat has been removed and its aorta and vena cava connected to the external blood circuit.

perfusion again resulted in the recovery of the original ATP level (Fig. 15e) but interestingly through the initial synthesis of ADP (Fig. 15d). In these spectra the signal-to-noise ratio is not particularly good as we were attempting to follow the time course of recovery. However once recovery was achieved we were able to keep the system in a steady state. We can then average the spectra over a longer period (Fig. 16) and demonstrate clearly that reperfusion after the second ischaemic period results in a nearly identical spectrum to that observed prior to ischaemia.

At this stage of the work we do not yet know which of the various parameters that we can measure (i.e. the depletion of the energy pool, intracellular pH, the ability of the kidney to recover its ATP level after reperfusion) is

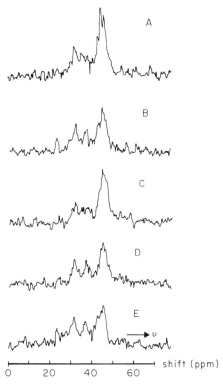

Fig. 15. ^{31}P NMR spectra of a single blood-perfused kidney. Data were collected under the same conditions as Fig. 13. The kidney was excised and rinsed with ice cold buffer. There was then a 15 min period of cold ischaemia and *spectrum A* was started together with perfusion. Data accumulation for *spectrum B* was initiated after 51 min of perfusion. After a total of 133 min the pump was turned off. *Spectrum C* was started after 12 min of *warm* ischaemia. Reperfusion began as data collection for *C* came to an end. *Spectra D* and *E* were begun after 12 and 24 min of reperfusion respectively.

important in relation to tissue viability. It is obvious however that ^{31}P NMR offers exciting new possibilities in studying some of the problems associated with transplant operations.

Summary and Perspective

We hope that in this presentation we have demonstrated that ^{31}P NMR can be used to study a range of biochemical, physiological and medical problems. The method has given a

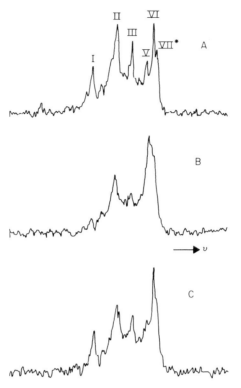

Fig. 16. Data were collected under the same conditions as Fig. 13. The kidney was excised, rinsed with ice cold buffer and then kept in the same buffer for 58 min. Data for *spectrum A* (800 scans) were collected 16 and 48 min after the start of perfusion. After 61 min of perfusion, the blood circuit was turned off. Data for the warm ischaemic kidney were collected for 23 min (*spectrum B*, 600 scans) and perfusion resumed immediately thereafter. *Spectrum C* shows data accumulated between 8 and 40 min after the onset of reperfusion.

new insight into the mechanism of energy dependent catecholamine accumulation into adreno-medullary storage vesicles. Evidence for metabolite compartmentation in skeletal muscle was presented. In perfused beating rat hearts, the function, injury and recovery can be followed. Finally, some problems associated with renal transplantation surgery can be and are being studied.

This conference is on 'NMR in Biology'. It is therefore appropriate to ask where our method provides new information that was previously not available from conventional

analytical techniques. The fact that the NMR method is
relatively rapid and non-destructive may be considered as
trivial by many. After all, you could argue that we have
known for a long time that the energy pool in a variety of
tissues is rapidly depleted under anoxic conditions. It is
true of course, that to establish, by conventional methods,
say the recovery of a tissue from an ischaemic insult,
would require measurements of many different samples after
destroying them. Therefore ^{31}P NMR can speed up the gather-
ing of information on the effect of different conditions
on tissue viability. No doubt, this is important, but can
only be done if ^{31}P NMR is linked up with other biochemical and physiological experiments. But NMR can tell us a
great deal about the environment of molecules in the intact tissue. Structures, intermolecular interactions and
spatial distributions can all be studied by NMR, as is
well documented in this volume. It is more than just a
hope that the same kind of detail can also be extracted
from NMR measurements on living systems. Phosphorus NMR
has particular advantages if one is interested in bioener-
getics but there is no reason why other nuclei should not
be used, as indeed is demonstrated in the two subsequent
papers.

Acknowledgements

A presentation like this that summarises work extending
over several years invariably has contributions from many
more people than the authors. Some of these are acknowledged by giving references to the appropriate papers. We
wish, however, to give particular thanks to Dr. David
Hoult, without whose expertise in electronic engineering,
instrument design and development this work would not have
been possible.

We also express thanks for the enthusiastic support
given by Dr. Rex Richards. The work on cardiac tissue has
been done in collaboration with Dr. Britton Chance and it

is a pleasure to acknowledge many years of happy and fruit-
ful collaboration with him. The work on kidney has been
done in collaboration with the Liverpool Transplant team
(see text). The production of this manuscript to meet the
deadline owes much to the enthusiasm of Mrs. Sue Bailey.

Financial support for the work was given by the Science
Research Council, for cardiac research by N.I.H. (3 PO1
HL18708-02S1 HEPA) and the British Heart Foundation. The
support by ICI Ltd. is also acknowledged.

References

1. Busby, S.J.W. and Radda, G.K. (1976). *In: Current Topics in Cellular Regulation*, eds. B.L. Horecker and E.R. Stadtman, Vol. X, Academic Press, New York, 89–160.
2. Gadian, D.G., Radda, G.K. and Richards, R.E. (1974). *Biochim. Biophys. Acta.* **358**, 57.
3. Hoult, D.I. and Richards, R.E. (1975). *Proc. Roy. Soc. Lond. Ser. A* **344**, 311–40.
4. Hoult, D.I., Busby, S.J.W., Gadian, D.G., Radda, G.K., Richards, R.E. and Seeley, P.J. (1974). *Nature* **252**, 285–7.
5. Busby, S.J.W., Gadian, D.G., Griffiths, J.R., Radda, G.K. and Richards, R.E. (1976). *Europ. J. Biochem.* **63**, 23–31.
6. Busby, S.J.W., Griffiths, J.R., Radda, G.K. and Seeley, P.J. (1975). *Proceedings of the Tenth FEBS Meeting*, North Holland, Vol. 40, pp. 75–85.
7. Salhany, J.M., Yamane, T.,Shulman, R.G. and Ogawa, S. (1975). *Proc. Nat. Acad. Sci. U.S.A.* **72**, 4966–70.
8. Dawson, J., Gadian, D.G. and Wilkie, D.R., this volume.
9. Busby, S.J.W. (1975). D.Phil. Thesis, Oxford.
10. Seeley, P.J. (1975). D.Phil. Thesis, Oxford.
11. Seeley, P.J., Busby, S.J.W., Gadian, D.G., Radda, G.K. and Richards, R.E. (1976). *Biochem. Soc. Trans.* **4**, 62–6.
12. Burt, C.T., Glonek, T. and Bárány, M. (1977). *Science* **195**, 145-9 (and references therein).
13. Dawson, J., Gadian, D.G. and Wilkie, D.R. (1977). *J. Physiol. (London)* **267**, part 3.
14. Burt, C.T., Glonek, T. and Bárány, M. (1976). *J. Biol. Chem.* **251**, 2584–91.
15. Gadian, D.G. (1974). D.Phil. Thesis, Oxford.
16. Busby, S.J.W., Gadian, D.G., Radda, G.K., Richards, R.E. and Seeley, P.J. (1977). Submitted for publication in *Biochem. J.*
17. Karpatkin, S. (1967). *J. Biol. Chem.* **242**, 3525–30.
18. Roodyn, D.B. (1965). *Int. Rev. Cytol.* **18**, 99–190.
19. Winkler, H. (1976). *Neuroscience* **1**, 65–80.
20. Casey, R.P., Njus, D., Radda, G.K., Seeley, P.J. and Sehr, P.A. (1977). *Horizons in Biochemistry* **3**, 224–52.
21. Bashford, C.L., Casey, R.P., Radda, G.K. and Ritchie, G.A. (1976). *Neuroscience* **1**, 399–412.

22. Bashford, C.L., Johnson, L.N., Radda, G.K. and Ritchie, G.A. (1976). *Europ. J. Biochem.* **67**, 105–14.
23. Bashford, C.L., Radda, G.K. and Ritchie, G.A. (1975). *FEBS Lett.* **50**, 21-4.
24. Bashford, C.L., Casey, R.P., Radda, G.K. and Ritchie, G.A. (1975). *Biochem. J.* **148**, 153–5.
25. Casey, R.P., Njus, D., Radda, G.K. and Sehr, P.A. (1977). *Biochemistry* **16**, 972–6.
26. Casey, R.P., Njus, D., Radda, G.K. and Sehr, P.A. (1976). *Biochem. J.* **158**, 583–8.
27. Radda, G.K. (1975). *Phil. Trans. R. Soc. Lond.* **B272**, 159–71.
28. Ritchie, G.A. (1975). D.Phil. Thesis, Oxford.
29. Douglas, W.W. (1968). *Br. J. Pharmacol.* **34**, 451–74.
30. Gadian, D.G., Hoult, D.I., Radda, G.K., Seeley, P.J., Chance, B. and Barlow, C. (1976). *Proc. Nat. Acad. Sci., U.S.A.* **73**, 4446–8.
31. Garlick, P.B., Radda, G.K., Seeley, P.J. and Chance, B. (1977). *Biochem. Biophys. Res. Comm.* **74**, 1256–62.
32. Jacobus, W.E., Taylor, G.J., Hollis, D.P. and Nunnally, R.L. (1977). *Nature* **265**, 756–8.
33. Radda, G.K., Sehr, P.A., Bore, P.J. and Sells, R.A. (1977). *Biochem. Biophys. Res. Comm.* (in press).

THE PROTON NMR SPECTRA OF WHOLE ORGANS

A.J. DANIELS,* J. KREBS, B.A. LEVINE,
P.E. WRIGHT and R.J.P. WILLIAMS

*Inorganic Chemistry and Molecular Biophysics Laboratories,
Oxford*

Introduction

The possibility that NMR spectroscopy could be used in
the study of whole organs seems to have occurred to many
people at about the same time. The earliest reference to
^{31}P spectroscopy was that of Moon and Richards [1] but
this was quickly followed by work at Oxford [2] and Chicago [2]. Recent work in Oxford is described in this
book. Carbon NMR spectroscopy of whole organs is described by Schaefer and coworkers [3] and ourselves [4] and
must have a big future once the incorporation of enriched
carbon has been thoroughly developed. The obvious advantage of plant materials cannot be overstressed. Proton
resonances are not so promising at first sight as protons
are too common in organic chemicals. However proton resonances have very different relaxation rates according to
the origin of the signal [4]. In slowly tumbling molecules, e.g. globular proteins of $\sim$ 100,000 molecular
weight, only a broad general mass of lines can be seen,
but in small molecules in solution resolution is no problem. Biological samples vary from solids to viscous solutions to almost free solutions containing both large and

*Present Address: Dept. Neurobiology, Catholic University of Chile,
Santiago, Chile.*

small molecules, so that there are a vast number of resonances which may be observed differentially using different
observational methods. In this paper we concentrate upon
the use of conventional high resolution F.T. methods such
as we have used in the examination of small molecules up
to molecular mass 50,000 daltons. We then see only molecules or fragments of molecules which have a fast tumbling
time.

An obvious starting point is to study those organs
which act as molecular storage centres. Table 1 lists some
that have been examined already, and some obvious attractive possibilities are brain systems, the organ of the
electric eel, slime mould colonies and glands of all
kinds. As ^{31}P studies have proved so successful in muscle
[2] we have no doubt that muscles can be analysed by ^{13}C
and ^{1}H once the technical problems are tackled. In this
paper we shall discuss our own observations on the materials in Table 1.

The Adrenal Gland

Even though the adrenal gland is a definite anatomical
unit in mammals, it consists of two different endocrine
organs: the adrenal cortex and the adrenal medulla. The
gland is surround by an elastic capsule which gives
hardly any ^{13}C, ^{31}P or ^{1}H spectrum when cleaned of con-
taminating cortex even though it contains many cells and
the materials of the membranes are present in high concentration. We note immediately that it is common experience that biological membranes give very poor ^{1}H NMR
spectra (^{31}P and ^{13}C spectra are better but still poor).

The Adrenal Cortex

The adrenal cortex gives a simple spectrum of unsaturated fats with very little evidence of the steroids stated
to be in the cortical zones (glomerulosa, fasciculata and
reticulosis) and stored in liposomes. In fact about 5 per

TABLE 1

Some Storage Systems in Biology open to NMR Study

Organ	Storage System	Spectral Observations (NMR)
Seeds Whole organ	Unsaturated Fatty Acids	^{13}C and ^1H successfully used
Zoo planckton Whole animal	Unsaturated Fatty Acids	^1H of fats seen but also some storage protein?
Adrenal Medulla Whole organ	ATP, Adrenaline and Protein	^1H, ^{13}C, ^{31}P observed
Adrenal Cortex	Sterol esters	Fatty chains of sterol esters observed by ^1H
Egg Yolk Vesicles	Phosvitin (protein)	^1H, ^{13}C, ^{31}P observed
Octopus Salivary Gland (Slices)	Neuro-transmitters and other unknown chemicals	$N^+(CH_3)_n$ observed (^1H) ? nucleic acid bases (^1H)
Splenic Nerve	Noradrenaline/ATP	^1H observed
Blood Platelets	Aromatic Molecules 5-Hydroxytrypt-amine/ATP	^1H observed in aromatic region

cent of the net weight of the adrenal cortex is cholesterol but in Fig. 1 there are no well resolved cholesterol resonances and the spectrum is largely that of the fatty chains of the sterol esters. No further information is obtained from ^{13}C or ^{31}P spectra. We conclude that the steroid esters are in a micelle in which the steroids are virtually immobilised but the long fatty acid chains are mobile [4].

The Adrenal Medulla

The adrenal medulla gives a rich spectrum. We observe the ^{31}P spectrum of ATP and the proton spectrum of ATP, adrenaline, and a protein in Fig. 2. After initial joint observations on ^{31}P this area of study has been continued by Dr. Richard's and Dr. Radda's group, while we have pursued the ^1H spectra. We note that the ^1H spectrum of the

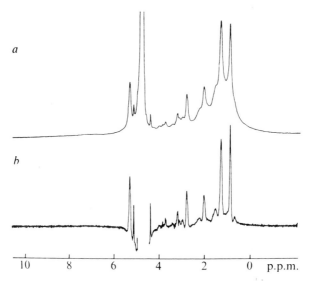

Fig. 1. NMR spectra of rat adrenal cortex (mainly the fasciculata and reticularis zones). *a*. Normal FT spectrum; *b*. Convolution difference spectrum showing a peak at 0.07 ppm assigned to the 18-methyl group of cholesterol.

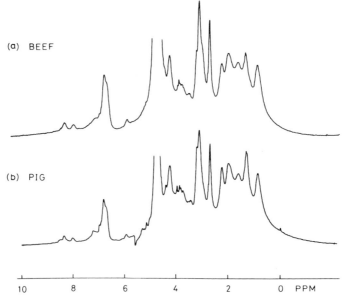

Fig. 2. The NMR spectrum of slices of Bovine and Pig adrenal glands showing from right to left peaks due to the chromogranin protein (1-2 ppm) aliphatic adrenaline peaks (around 3 ppm) aromatic adrenaline peaks (6.7 ppm) and ATP peaks (5.8, 8.0 and 8.3 ppm).

two main components in the medulla is not that of free
ATP and adrenaline at low concentration. This applies whether we look at whole glands (rats), gland slices (pig) or
the isolated hormone storage vesicles. The fact that the
vesicles have the same spectrum after preparation as in
the freshly dissected whole gland proves for the first
time that the vesicles are not damaged by isolation and
hence are true representatives of the situation inside
the cell. NMR spectroscopy is the only technique which can
do this. The preparation of artificial mixtures of adrenaline, ATP, and soluble proteins (at concentrations close
to the ones inside chromaffin vesicles) gives a very similar spectrum to that of the vesicles. Thus we can determine the ring current shifts on each resonance and later
we shall use them for structural purposes, see below.

The major component of the soluble proteins, Chromogranin A, can be isolated and purified. This protein was
known to be in a random coil configuration *in vitro* [5,6].
The NMR spectrum of the protein in the vesicle (*in vivo*)
and the spectrum of the purified protein (*in vitro*) are
closely similar and they are both similar to a random coil
synthesised spectrum. The spectra are all high resolution
spectra despite the fact that the protein has a molecular
weight of 80,000. The T_1 and T_2 relaxation times are also
such as to indicate that the protein has a random configuration. We conclude that the vesicles *in vivo* contain a
large random coil protein in what is not a very viscous
solution.

Further details of the structure can be obtained by
using isomorphous substitution of the Mg^{2+} and Ca^{2+} ions,
known to be in the vesicle, by Mn^{2+}. The substitution can
be made by exposing gland slices or isolated vesicles to
Mn^{2+} solutions. The uptake of the cation causes a broadening of resonances close to its binding site. Our results
show a broadening of some of the resonances of ATP, some
resonances of adrenaline and some broadening of the gluta-

mic acid resonances of the protein. (Note the protein is very acidic and rich in glutamic acid.) A model study of Gd(III) ATP plus adrenaline gave very similar relative broadenings of resonances. From these data and the ring current shifts we can construct a structural unit which must be quite highly populated by the ATP and adrenaline molecules of the vesicle.

Now we see the functional significance of the vesicle contents. Each component helps to lower the osmotic pressure of another through mutual association and cross-linking in short chains of the type protein (glutamate): metal ion or adrenaline: ATP: adrenaline: protein etc. These chains are branched and the whole contents are contained in a loose jelly. The random-coil acidic protein fulfils a major role of stabilising the jelly and preventing precipitation.

The uptake of manganese may have two additional importances. Manganese is known to be a neural poison with important implications in brain amine levels [7]. We can suggest from the above work that manganese could act on the nerve terminals through binding in their vesicles.

Equally useful however the above study shows how an analysis of compartments in biology can be made using relaxation (or shift) probes. This applies to ^{1}H, ^{13}C or ^{31}P studies. We have not observed in our work any evidence for separate populations of adrenaline in the storage vesicle of the adrenal gland.

Apart from the study of the storage system we can follow release of its contents down to the level of approximately 1 per cent. This is possible as the released adrenaline gives *sharp* peaks at different fields from the bound adrenaline peaks. Again as the relaxation times of free and bound molecules are very different we can clearly separate them using pulse methods. (Note that pulse methods in general can be used in spectral simplification too.)

On the suggestion of Dr. J. Costa of N.I.H. (Bethseda U.S.A.) we have tried to incorporate fluorine substituted amines into the adrenal vesicles. The results have not been very promising so far but the method should be closely examined. Successful introduction of a fluorine atom gives a non-natural probe in any molecule of interest and the possibilities of extracting information are therefore very large. Mechanisms of action of drugs could be studied using this method.

A Zooplanckton

The food chain of the oceans is phytoplanckton (hv + CO_2 etc) → zooplanckton → fish. The zooplanckton must over-winter in the North Atlantic without a food supply and they do so by building up huge stores of fatty acids. A brief study we have made indicates that the synthetic pathway might be followed *in vivo* by ^{13}C NMR. The procedure would follow that of Schaefer on seed growth [3]. At the present time we have only looked in detail at the ^{1}H spectrum of the whole organism. This could be called a feasibility study. We observe very easily the stores of *unsaturated* fats but we also noticed a store of some aromatic material. Analysis of the material by conventional biochemical methods indicates that it is a protein. We are led to ask the question as to whether this organism has a store of protein (nitrogen-store) as well as of fat (carbon and hydrogen). The work is in its early stages but one could imagine a study in which the life cycle of the animal was followed in the NMR tube, and stretching the imagination, there is no reason why a colony of phyto- and zoo-planckton fed with $^{13}CO_2$ could not be examined.

Phosphorus Storage

The protein phosvitin occurs in many eggs of birds and insects as a storage form of phosphorus in egg-yolk vesicles. It is a large protein containing many phospho-

serines. The ^{31}P resonance is readily observed in the
vesicles as well as both the ^{13}C and ^{1}H resonances. The
protein is largely random coil, strung out in this
fashion, no doubt, by the negative charge repulsion. It
has some structured regions of course. The protein has
other interesting features for unlike most proteins its
histidines are locked in by a field of negatively charged
phosphates. As both phosphate and histidine have a pK_a of
about 6.5 the histidine and phosphorus pH titrations are
quite unusual. The ionisation state of these groups will
control the conformation of the protein of course.

We draw attention to the work of Colman and Gadian [8]
on the metabolism of phosphorus in the developing tadpole
as studied by NMR.

The Salivary Gland of the Octopus

This gland is not just a digestive unit but contains
neurologically active poisons so that when an octopus
bites its prey it paralizes it. Digestion follows. The
study of this organ as a slice in a fresh state by ^{1}H
NMR revealed intense signals of $\overset{+}{N}$-$(CH_3)_n$ groups which
would not appear to be accompanied by -CH_2-$\overset{+}{N}$ resonances.
Of great surprise however was the appearance of resonances
beyond the aromatic region at ~8.7 ppm. These resonances
are strong so that the material is > 2mM in the whole
organ. The resonances must belong to a nitrogen-
substituted ring system of a nuclei acid base or a unit
such as nicotinamide. The group has a pK_a < 2 but no pK_a
between 2 and 10. It is not bound to a sugar moiety — no
H(1') resonance. What is this heterocyclic base and what
is it for? (Note it is probably not adenine.)

In other regions of the spectrum we see signals but
their intensity depends on sample and sample age. In the
case of old samples the ^{1}H signals of virtually all amino-
acids are observable at their correct free amino-acid
positions, Fig. 3. Undoubtedly we are following the onset

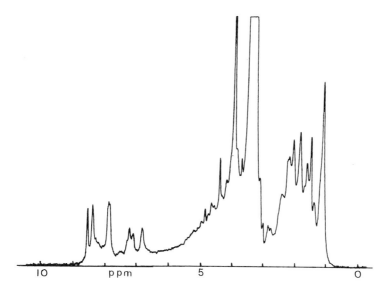

Fig. 3. The new spectrum of the salivary gland of the octopus after standing for some hours shows in addition to peaks due to N-CH$_3$ groups and peaks at >8.0 ppm the peaks of free amino acids arising from proteolysis.

of self-proteolysis by the highly active proteases of the salivary gland. This illustrates the kinetic use of NMR in whole organs incidentally. There are other signals in the aliphatic region at ~2.5 and 3.0 ppm which cannot be ascribed to any particular groups as yet.

Brain Slices

As this examination of the use of ^{1}H NMR in the study of whole organs is really a test exercise we have extended our study to casual observation of material from the central and peripheric nervous system. When beef splenic nerve axons are examined a remarkably well resolved spectrum is obtained from the preparation particularly in the aromatic region of the spectrum where we can see resonances tentatively assigned to adenine nucleotides and hydroxyphenols.

In all brain samples examined (corpus striatum, hypothalamus, corpus callosum) we find strong signals from nucleotide bases at around 8.0 ± 3.0 ppm. This observation will be followed up by a comparative study of a large number of organs.

Cell Colonies

The work we have done to date on the NMR of whole organs is a screening of many different organs and the detailed study of one. We are looking at static states of given systems but we know we can move on to their dynamic states for in the very first experiments we could follow release of adrenaline from glands as a sharp discrete *separate* signal [2]. Apart from characterising the chemicals in an organ we can also obtain structural information about molecular assemblies and compartments in space [2].

Beyond this situation we have also followed the metabolism of protein digestion using the octopus salivary gland. This experiment proves that it will be possible to follow metabolic events in many organisms. In trial experiments Dr. Dobson has found that he can look at colonies of cells from slime mould and yeast. There is no reason to suppose that this work cannot be extended to cells in culture in order to study colonies of diseased cells as compared with normal cells. The work of the American group on seeds shows how the metabolism of plants can be done using ^{13}C NMR and there are obvious possible extensions to growing animal cells using ^{13}C enriched sugars with culture cells or organs. There is a place too for ^{19}F NMR in uptake experiments.

Acknowledgements

We thank Dr. M. Fillenz for her considerable assistance. R.J.P.W. is a member of the Oxford Enzyme Group. The work is supported by M.R.C., S.R.C. and the Wellcome Trust.

References

1. Moon, R.B. and Richards, J.H. (1973). *J. Biol. Chem.* **248**, 7276.
2. Hoult, D.I., Busby, S.J.W., Gadian, D.G., Radda, G.K., Richards, R.E. and Seeley, P.J. (1974). *Nature* **252**, 285.
   Daniels, A., Korda, A., Tanswell, P., Williams, A. and Williams, R.J.P. (1974). *Proc. Roy. Soc.* **B187**, 353.
   Henderson, T.O., Costello, A.J.R. and Amachi, A. (1974). *Proc. Nat. Acad. Sci. U.S.A.* **71**, 2487.
3. Schaefer, J., Stejskal, E.O. and Beard, C.F. (1975). *Plant Physiol.* **55**, 1048.
4. Daniels, A., Williams, R.J.P. and Wright, P.E. (1976). *Nature* **261**, 321.
5. Smith, A.D. and Winkler, H. (1976). *Biochem. J.* **103**, 480 and 483.
6. Kirshner, H.G. and Kirshner, N. (1969). *Biochem. Biophys. Acta* **181**, 219.
7. Cotzias, G.C. (1973). *In: The Harvey Lectures Series* **68**, 115. Academic Press, London.
8. Colman, A. and Gadian, D.G. (1976). *Eur. J. Biochem.* **61**, 387.

STUDIES OF LIVING, CONTRACTING MUSCLE BY ^{31}P NUCLEAR MAGNETIC RESONANCE

JOAN DAWSON, D.G. GADIAN* and D.R. WILKIE

Dept. of Physiology, University College, London

Introduction

The State of the Art

The time has passed when news was made simply by obtaining spectra from pieces of intact or semi-intact tissue. In the case of muscle this was already accomplished by 1974, when the group at Oxford [1] reported that resonances could be obtained representing the important phosphorus metabolites in whole rat muscles. Since that time a number of ^{31}P NMR studies on whole muscle or pieces of muscle have been reported from different laboratories [2,3,4,5].

In all of these studies the specimen tube was simply packed with muscles at around room temperature, under which circumstances they died fairly rapidly from anoxia (see [1], Figs. 3 and 4). Nevertheless a very useful body of information has been built up, upon which there is gratifying agreement between different laboratories. The chemical shifts reported for different compounds are very consistent and the quantitative estimates for these compounds agree reasonably with direct chemical analysis. Unexpected resonances are frequently observed in the phosphodiester region of the spectrum which are surpris-

─────────────

Dept. of Biochemistry, South Parks Road, Oxford.

ingly large in semitendinosus muscle from the rabbit [4]
and in toad gastrocnemius muscle [3].

An important contribution of the studies cited above
was to show that it was worth the effort required to develop the techniques necessary to maintain muscles in
physiological condition within the spectrometer so that
^{31}P NMR could become a useful tool in the study of living,
contracting muscle. Our success in this effort was reported to the Physiological Society a year before the present
Conference [6]; since that time we have used ^{31}P NMR in
an extensive study of the biochemical changes occurring
during contraction and the following metabolic recovery
period [7].

*Advantages and Limitations of ^{31}P NMR for the study of
intact tissues*

Many of the metabolites known to be involved in cellular chemical energy transfers contain phosphorus and
occur free in solution at sufficiently high concentrations
(about 0.5 mmol kg^{-1}) to be detectable by ^{31}P NMR. In
studying the reactions that these compounds undergo as a
result of physiological activity ^{31}P NMR has two specific
advantages not offered by ordinary chemical methods:

Firstly, it is *noninvasive* and thus allows the time-
course of metabolic reactions to be followed in a single
experiment on a single preparation, such as a set of
muscles. This is to be contrasted with chemical methods
which require freezing the tissue to stop chemical reactions at some particular time during or following activity. Different specimens must be used for each point in
time and enough experiments must be done for statistical
analysis. All this involves many weeks of work, and in
studies of muscle gives rise to serious problems from
variation between batches of experimental animals [8].

Because NMR is noninvasive it also provides information
about the genuine intracellular environment of metabolites

which is lost if the tissue must be destroyed for chemical analysis. For example, ^{31}P NMR has given information about internal pH and about the binding of Mg^{++} to ATP. The technique also suggests that the intracellular environment may not be homogenous throughout the tissue but may contain compartments that differ significantly in pH and in other respects [9,10].

The second major advantage of ^{31}P NMR is that it is *nonspecific*; resonances are observed from all freely mobile phosphorus compounds present in high enough concentrations. The simultaneous observation of all these compounds, rather than just those selected for chemical analysis could permit detection of phosphorus metabolism not predicted by current theories and which might be missed by conventional methods. Already resonances have been observed from several phosphodiester compounds of unknown functional significance, only one of which has yet been chemically identified. ^{31}P NMR allows us to investigate the function of these compounds, by looking for changes in the size or in other characteristics of the resonances they produce under various conditions, even before the compounds themselves have been chemically identified.

The main disadvantage of ^{31}P NMR in the study of intact tissue is the familiar one of low sensitivity compared with chemical analysis, coupled with the fact that metabolites are present in relatively low concentration. For this reason all the successful studies of intact tissue have employed sophisticated signal detection and analysis systems.

The Study of Muscular Contraction

The problem of low sensitivity is exacerbated if one wishes to study the biochemical reactions associated with muscular activity, since the reactions during contraction itself are extremely rapid (typically lasting a few

seconds) and those during recovery are fairly rapid (typically lasting a few minutes). In order to obtain sufficient
signal it is necessary to average many scans, and these can
be repeated only every 2 sec because it takes a relatively
long time (determined by T_1, the spin-lattice relaxation
time) for the ^{31}P nuclei to return to their unexcited
state. As a result, the accumulation of a spectrum can take
several hours.

Since the muscles must be kept in a steady-state at
rest or making a series of contractions throughout the
experiment, severe constraints are imposed upon the choice
of preparation and experimental conditions. The most ele-
mentary requirements are that the muscles be maintained
in a physiological salt solution (Ringer's solution) with
an ionic composition similar to that of blood, and that
they be adequately oxygenated. In our experiments oxygenated Ringer's solution was continuously perfused through
the sample tube. Once a physiological steady-state is
achieved, the inherent slowness of the method can be overcome by stimulating the muscles repeatedly and obtaining
scans at predetermined times after the beginning of the
stimulation. Thus, spectra can be accumulated representing
the biochemical situation during the actual contraction
and during specified intervals throughout the recovery
period.

The possibility of using a totally new technique in the
study of the biochemical reactions associated with contraction is particularly welcome at the present time. The
methods currently used to study living muscle, ultra-
rapid freezing at some instant during or following a contraction [11] followed by extraction and chemical analysis,
have given results which indicate a quantitative discrepancy between the amount of mechanical and thermal energy
produced by the muscle and the amount of chemical fuel
consumed. This discrepancy suggests that purely chemical
methods may have missed some important reaction [8].

The actual fuel which powers contraction is adenosine triphosphate (ATP) whose hydrolysis to adenosine diphosphate (ADP) and inorganic phosphate (Pi) accompanies the development of mechanical energy during contraction. However, the concentration of ATP in the muscle does not change during a normal contraction because it is rapidly rephosphorylated at the expense of phosphocreatine (PCr) by the enzyme creatine transphosphorylase. Following contraction the PCr is rebuilt to resting levels as a result of the breakdown of glycogen in a series to steps to form lactic acid (in the absence of oxygen) or CO_2 and H_2O (if oxygen is present).

A general description of the biochemical events associated with contraction is published elsewhere [12]. Since these glycolytic and oxidative reactions begin relatively slowly it is possible to design experiments in which the hydrolysis of PCr is the only identifiable net chemical reaction proceeding during contraction. Under these conditions it is always found that the breakdown of PCr is insufficient (only half enough in brief contractions) to account for the concurrent production of heat and work [8]; for later references see [13,14,15,16]; and reviews [17] and [18]. The hydrolysis of PCr during contraction is also smaller than would be predicted by current biochemical theories from the subsequent metabolic recovery processes, whether these occur aerobically and are measured by oxygen consumption [19] or take place in the absence of oxygen and are measured by lactate production [20].

Since muscle, like any other machine, must obey the first law of thermodynamics, the highly significant imbalance in the energy equation indicates a significant gap in our knowledge of the biochemistry of contracting muscle. Quite possibly the availability of a completely novel technique might resolve this discrepancy. Whether or not this particular problem is solved by ^{31}P NMR, the technique is, in fact, of general utility in studying muscle metabolism.

NMR is a complicated and expensive technique. In the
following section we present details of the methods we
have devised for using ^{31}P NMR in the study of living
muscle. The requirements are exacting and no doubt will
continue to be so even as greater experience is gained.
There is little point, however, in using this powerful
physical technique to do biochemical or physiological
experiments unless appropriate attention is paid to the
biological part of the problem. This includes both choos-
ing the questions which are best suited to study by ^{31}P
NMR and careful attention to the adequacy of the biological preparation used to answer them.

Methods

The Choice of Experimental Preparation

This choice involves a compromise. Due to the insensi-
tivity of the technique one wishes to work with as large
a sample, and thus as large a signal as possible. Since
the time required to achieve a given signal-to-noise ratio
increases as the square of the sample cross-section, there
is much advantage in filling the sample volume completely
with tissue. This was, in fact, done in the earlier ^{31}P
NMR studies (see Introduction). However, in order to keep
the muscle in a physiological condition, it must be adequately oxygenated. In the experiments reported here oxygen was delivered by diffusion from the bathing medium
surrounding the muscles; this can only be accomplished
with very thin muscles separated by an appreciable volume
of fluid. We have not yet attempted to perfuse small
skeletal muscles through their own circulartory system
though we intend to do this soon. In addition it is highly
desirable to choose muscles which function at low temperatures since the rates of biochemical processes, including
oxygen consumption, are then decreased [21].

We have performed many of our experiments on frog
sartorii from *Rana temporaria* at 4°C. These muscles are

only about 1 mm thick (by approximately 5 mm wide and 30
mm long) so it is relatively easy to supply them with oxygen by diffusion. 4°C was chosen rather than 0°C, which
is usual in physiological experiments on amphibian muscles,
in order to provide a margin of safety against freezing.
Even though we are able to put 4 frog sartorii in the
sample tube, only about a third of the sample cross-section
is occupied by muscle. For certain experiments, therefore,
we used pairs of small gastrocnemii, either from *Rana*
temporaria or from the toad *Bufo bufo*. These muscles are
roughly circular in cross-section, the ones we used having
a diameter of about 4 mm at the widest point and weighing
about 100 mg each. These gastrocnemii fill up appreciably
more of the sample space, but are not as well oxygenated.
In our latest experiments, in which we have been studying
metabolism under anaerobic conditions, we have been able
to use larger gastrocnemii weighing 400--500 mg each.

Physiological Techniques

For a fuller discussion of the physiological techniques,
see Dawson *et al*. [7].
Design of chamber The glass NMR sample tube, 7.5 mm in
diameter and 70 mm long was converted into an experimental chamber, as shown in Fig. 1. The support system is constructed entirely of teflon, glass and epoxy resin as these
materials do not appreciably disturb either the homogeneity of the magnetic field or the tuning of the RF coils.

The muscle holder consists of a central glass capillary
tube to which are attached two teflon bobbins. One of these
serves as the top to the experimental chamber, and the
other, a circular disc with two slots cut out, is fixed
to the capillary tube approximately 1 mm from the bottom.
The ends of the muscles are inserted into these slots so
that small attached pieces of bone form stops below the
bobbin. Cotton tied to the tendon at the other end of the
muscle is threaded through holes in the top of the chamber

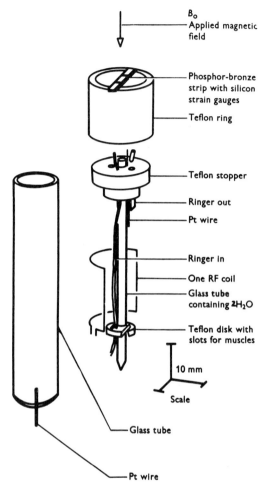

Fig. 1. Design of experimental chamber. The arrangements for stimulating, recording tension and perfusing oxygenated Ringer's solution are
further described in the text. The volume in which the NMR measurement
is made is defined by the two single-turn RF coils, only one of which
is shown in the diagram. They extend from 3 mm to 18 mm above the top
of the lower teflon disc.

and tied over the force transducer which is fitted onto
another teflon support above. The muscles are held in a
vertical position parallel to the central glass tube and
the magnetic field. The capillary tube serves an additional purpose, being filled with a solution of KCl in

D_2O. The KC1 (134g/kg D_2O) is present to prevent the D_2O from freezing at the temperatures of around 4°C at which the experiments are conducted (the freezing point of pure D_2O is 3.81°C). The resonance from deuterium serves by a feedback arrangement to hold the magnetic field steady to about one part in 10^8, and to optimize the spatial homogeneity of the field.

Stimulation and recording The muscles are stimulated electrically via two axially located platinum wires, one threaded through the teflon top and one sealed into the bottom of the chamber, and the resulting force development is recorded by a transducer consisting of two silicon strain gauges forming a bridge circuit bonded above and below a narrow phosphor-bronze strip attached across a circular teflon support (see Fig. 1).

Perfusion of oxygenated Ringer's solution Bubbling air or oxygen through the experimental chamber would result in severe disturbance of magnetic homogeneity; therefore, a recirculating perfusion system was chosen as the method of oxygenation. Approximately 200 ml of Ringer's solution (composition in mmol 1^{-1}: NaCl 111; KC1 2.5; $CaCl_2$ 2; tris [tri(hydroxymethyl)methylamine] 10, adjusted to pH 7.2 at 25°C with HC1) was bubbled with 100 per cent oxygen at room temperature. This was pumped to and from the chamber with a Pharmacia P3 variable speed peristaltic pump. The pH increases to 7.7 when the temperature falls to 4°C.

Preparation of muscle extracts Extracts were made by soaking the frozen muscles for four days in 3 ml of 1.25 mmol 1^{-1} EDTA (ethylene diamine tetra-acetic acid) in 50 per cent methanol, pH 7.5, at -30°C. The fluid was freeze-dried and redissolved in the required volume of distilled water.

NMR Spectroscopy

^{31}P NMR spectra were recorded at 129.2 MHz on a spectrometer constructed in the Oxford laboratory and interfaced

with a Nicolet B-NC 12 computer. Details of this spectrometer have been reported by Hoult and Richards [22]. The
spectrometer was operated in the Fourier Transform mode
and employed a deuterium field-frequency lock. Spectra
were accumulated for periods of up to 10 hours and stored
on a magnetic disc. In high resolution NMR the sample is
normally spun to improve the resolution, but unfortunately
in our experiments this was impossible.

Identification of resonances and pH determination   Identification of the muscle resonances is made by comparing
the resonance positions (chemical shifts) with those of
metabolites free in solution. In this way the resonances
of ATP, PCr, Pi and sugar phosphates (Sugar P) can be
unequivocally assigned [1,2,7]. The position of the PCr
resonance has been shown to be very stable in intact
muscle (see [7]); we have therefore measured all our chemical shifts relative to that of the PCr resonance. The
chemical shift of the Pi resonance, on the other hand, is
extremely sensitive to pH in the region of neutrality.
Thus, after construction of a suitable pH titration curve
for the Pi resonance in a solution of physiological ionic
strength, it is possible to determine the internal pH of
the muscle fibres directly from the chemical shift difference between the PCr and Pi resonances.

Determinations of concentration   The areas of the various
resonances are proportional to the quantities of metabolites within the volume enclosed by the RF coils, provided
that the RF pulses are applied at time intervals much
greater than the spin-lattice relaxation time, T_1, of the
resonances. However, in order to optimize the signal-to-
noise ratio of the spectra, we normally applied 70° pulses
at intervals of 2 sec, which approximates to the T_1 values
of the resonances. Under these conditions the areas of the
resonances are reduced, an effect known as 'saturation',
by factors determined by the T_1 values. We have determined
these factors in appropriate experiments on frog gastroc-

nemius muscles (see [7]). In experiments in which scans
are repeated every 2 sec, the concentration of ATP, PCr,
sugar P, Pi and the unidentified resonances around -3 ppm
are arrived at by multiplying their respective resonance
areas by 1.0, 1.25, 1.5, 1.8 and 1.3, respectively.

Quantitative calibration of metabolite levels in absolute terms is complicated by the facts that the exact
volume that contributes signal cannot be determined and
that the sample contains both muscle tissue and an unknown
amount of circulating Ringer's solution. For these reasons
a special calibration method was employed (see [7]).

This calibration procedure is expensive in machine time
and cannot be repeated in each experiment. For many purposes the ratio of peak areas provides a sufficient basis
for interpretation.

Control of timing A computer program was written to syn-
chronise the accumulation of NMR data with the electrical
stimulation of the muscle. A given experiment begins at
time 0 with a pulse from the computer which, after a preset delay (typically 1 sec), triggers the stimulator and
causes the muscle to contract. The RF pulses commence 2
sec from time 0 and are repeated every 2 sec throughout
the course of the experiment. The first m scans are stored
in Bin 1, the next m scans in Bin 2, etc., until n bins
(mn scans) have been accumulated. A trigger pulse is then
provided and the cycle recommences. The process is repeated many times to built up the required signal-to-noise
ratio. By this means, although experiments may last many
hours, the kinetics of reactions can be followed with a
time resolution set by the minimal interval between successive bins, that is 2 sec.

For example, in the experiments involving 25 sec stimulations, m = 200 and n = 8; so that 8 bins, each of period
400 sec, are collected. The time interval between successive tetani is 3200 sec. The program can be suitably modified for other patterns of stimulation.

Results

Resting Muscle

Fig. 2 shows ^{31}P NMR spectra obtained on well-oxygenated
frog sartorius (2A) and toad gastrocnemius (2B) muscles
maintained at 4°C. These spectra differ from those obtained
in previous studies of intact muscle because in the present
case the muscles were maintained in a physiological condition, a fact which is clearly illustrated by the relative
amounts of metabolites present. The PCr resonances are
large in these spectra and those of Pi are small; there is
so little sugar P that one cannot be sure that peaks representing these compounds are present. Conditions such as
anoxia which lead to muscle deterioration all too easily
cause the progressively increasing levels of Pi and sugar
P which were characteristic of earlier ^{31}P NMR studies of
intact muscle. In the spectra shown in Fig. 2 the positions
of the Pi resonances indicate a pH of 7.5 fcr the frog sartorius and 7.4 for the toad gastrocnemius. Factors influencing pH in excised muscles will be dealt with in the
Discussion.

The spectrum of frog sartorius muscle shown in Fig. 2a
was calibrated for concentrations of metabolites by a method described elsewhere (see [7]). The muscles were then
frozen in liquid nitrogen and their methanol extracts were
analysed chemically. The PCr content was determined by ^{31}P
NMR to be 23.5 mmol 1^{-1} as compared to 22.5 mmol 1^{-1} obtained by chemical analysis with correction for density
and extracellular space. We have also compared the ratios
of PCr/ATP and PCr/Pi determined by ^{31}P NMR in several
resting muscles with the ratios of these compounds obtained
during several studies by conventional methods in the London laboratory. NMR tends to give a slightly lower estimate
of PCr/ATP and a slightly higher one for PCr/Pi, but the
differences are within experimental error, due to the
large standard deviation associated with the chemical

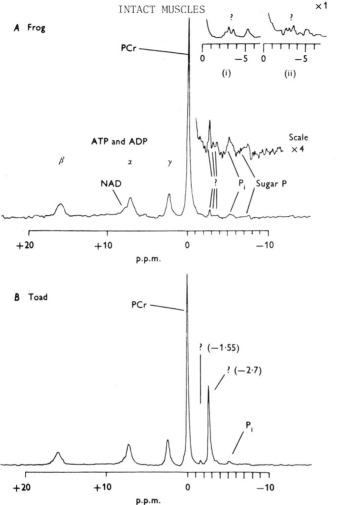

Fig. 2. Spectra from resting muscles at 4°C. Ordinates: signal. Abscissae: chemical shift in ppm; the frequency increases from left to right, the total span representing 5 kHz. A. 4 frog sartorii, May 19, 1976. Lowest line: average of 10,000 scans at 2 sec intervals, with peaks assigned as described in text. The short length of record on the right hand side of the PCr peak has been enlarged 4x vertically to show fine detail. Note that of the three unidentified peaks at -2.7, -3.05 and -3.6 ppm, only the left hand one is conspicuous. The inset figs show how these peaks vary in size, though not in position. (i) 16 November, 1975; (ii) 29 April, 1976. B. 2 toad gastrocnemii, May 20, 1976. Average of 6,000 scans at 4 sec intervals.

methods (for a table of these results, see [7]). Since values for the concentrations of known metabolites that are obtained by ^{31}P NMR are very similar to those obtained

by chemical analysis, we conclude that each of these two
entirely different techniques tends to confirm the reli-
ability of the other.

Fig. 2 confirms the observation in earlier ^{31}P NMR
studies that several previously unexpected resonances are
present in spectra from intact muscle. In Fig. 2A these
are the peaks at -2.7, -3.05 and -3.6 ppm. The same three
resonances are observed in spectra from toad gastrocnemius
muscles, but in this tissue the resonance at -2.7 ppm is
extremely large, as first shown by Burt and coworkers [3].
In Fig. 2B the unidentified resonance at -2.7 ppm is almost
half as big as that of PCr. Fig. 2B also shows a small but
definite peak at -1.55 ppm; this appears to be characteristic of toad muscle, and was observed for the first time
in the present study due to the relatively good signal-to-
noise ratio.

Spectra from methanol extracts of frog and toad muscle
are in every way consistent with those from intact muscle.
Of particular interest is the fact that all of the resonances observed in the region between -1.5 and -3.6 ppm in
spectra from intact muscle are also present in the same
relative sizes in spectra from extracts of the same type
of muscle. Thus, each of these resonances must arise from
a different compound, rather than from the same compound
giving rise to two or more resonances as a result of compartmentation in intact muscle.

Stimulated Muscles

Effect of a single prolonged contraction Large changes in
metabolite concentrations occur after prolonged stimulation. This is demonstrated in Fig. 3, which shows (A) a
spectrum of a pair of resting toad gastrocnemius muscles,
and (B) the spectrum of the same muscles after a single
tetanic stimulation lasting 35 sec. In this experiment
recovery was impeded by turning off the perfusion pump.
The changes are as expected; PCr falls to about half its

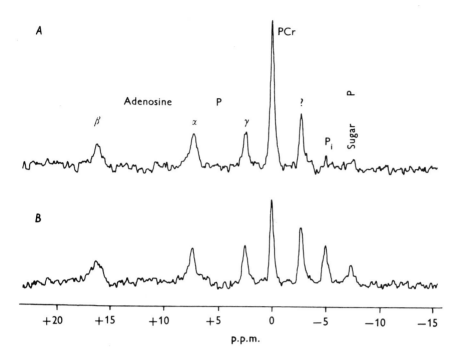

Fig. 3. The effect or prolonged stimulation on the ^{31}P spectrum of
toad gastrocnemius. Spectrum A was obtained on a pair of small toad
gastrocnemii in oxygenated Ringer's solution at 4°C. The muscles were
then stimulated for 35 sec, during which time the tension fell from
1.58 N to 0.63 N. The perfusion was turned off to retard recovery and
spectrum B was accumulated. Each spectrum was obtained over a period
of 7 min, and represents an accumulation of 200 scans at 2 sec intervals.

resting level, and there are large increases in the amounts
amounts of Pi and sugar P. The ATP level remains constant.
The initial pH indicated by the Pi resonance is 7.2; this
does not change during the 7 min after relaxation.

The changes demonstrated by Fig. 3 are very dramatic,
and confirm that it is possible to produce large metabolic
responses to electrical stimulation and to measure them in
a single experiment using ^{31}P NMR. It is, however, more
interesting to study the metabolic changes associated with
contractions from which the muscles actually recover.
Contraction and recovery Two different patterns of muscle
stimulation were employed: 25 sec stimulations every 56

min were used to study recovery after large metabolic changes, whereas 1 sec stimulations every 2 min were used when accumulating spectra during contraction.

Each of these patterns is very near the maximum that the muscles can sustain without severe progressive deterioration. The muscles are stimulated for almost the same number of seconds per hour in the two cases; thus, any differences in the metabolic recovery processes must result from the different patterns of stimulation, rather than from different amounts of activity. Sartorius and gastrocnemius muscles of the frog, and gastrocnemius muscles of the toad were used in both types of experiment. *Recovery after long contractions* Frog sartorii were stimulated for 25 sec every 56 min and scans from the recovery period between contractions were accumulated in 8 bins, each of 7 min duration. Spectra from the first bin (Fig. 4A) and the last 4 bins (Fig. 4B) are shown, and Fig. 4C shows the variations in metabolite concentration with time after contraction.

There is a clear difference in concentration of PCr between the two spectra shown in Fig. 4, indicating a large breakdown of this compound during contraction, with subsequent regeneration. The rebuilding of PCr is roughly exponential, with a half-time of about 10 min; extrapolation to zero time indicates that 20 per cent of the PCr is broken down during contraction. The concentration of Pi is increased about 4-fold immediately after contraction, which is roughly equivalent to the phosphate released upon hydrolysis of PCr. The half-time of removal of Pi corresponds with that of the rebuilding of PCr. Interestingly, during recovery the Pi falls to a level which is significantly lower than that observed in resting, unstimulated muscles.

The amount of sugar P increases greatly after these long contractions, and remains high throughout the recovery period. This is shown by both the spectra of Fig. 4, and

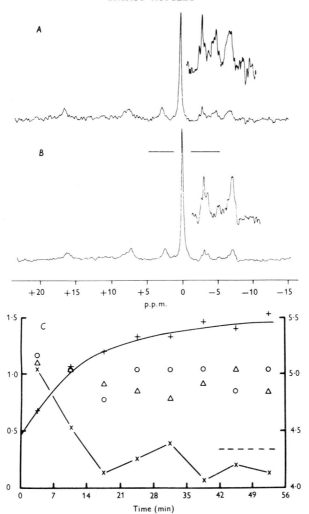

Fig. 4. Recovery of frog sartorius from long contractions.   Four frog
sartorius muscles were repeatedly stimulated for 25 sec every 56 min
and spectra accumulated into 8 bins of 7 min each. Spectrum A was
accumulated from 0-7 min after the contraction and spectrum B was
obtained throughout the last 28 min of recovery. The horizontal line
at the PCr resonance in spectrum B indicates the height of this peak
in spectrum A. Scans were made at 2 sec intervals. The muscles gave
7 virtually identical responses, during each of which the tension fell
from 1.22 N to 0.47 N. The graph (C) shows how PCr +, ATP O, Pi x and
Sugar P Δ varied throughout the 8 bins. The ordinates show the resonance peak areas as multiples of the mean area for the β ATP peak. The
right-hand scale applies to PCr. The exponential curve drawn through
the PCr points has a $T_{\frac{1}{2}}$ of 9.1 min.

Fig. 4C shows a high and constant level of these compounds,
approximately 6 mmol l^{-1}. In contrast, very little sugar P
is detectable in resting muscle (see Fig. 2). In one experiment a spectrum was obtained six hours after the last
stimulation, at which time the sugar P had fallen by 70
per cent, showing that these changes can be reversed in
oxygenated muscle. The spectra shown in Fig. 4 were obtained by adding scans accumulated after 7 contractions of
the same set of sartorius muscles; it is therefore not
clear to what extent the sugar P rose after each contraction. In a later experiment four small frog gastrocnemius
muscles weighing about 100 mg each were placed together in
the chamber, and sufficient signal-to-noise ratio was thus
achieved to accumulate spectra representing each minute
following a single 15 sec contraction. This experiment
indicated that a large amount of sugar P was formed immediately after the first of a series of contractions, but
that subsequent contractions generated relatively little
sugar P.

Short contractions: During contraction The spectra accumulated during the actual contractions of frog sartorius
or toad gastrocnemius stimulated for 1 sec every 2 mins
were not different in any apparent way from those accumulated during the recovery interval between contractions
(see [7], Fig. 8). Only a moderate signal-to-noise ratio
was achieved in these experiments, because only one scan
can be made during each contraction, but the results do
rule out any *dramatic* changes in concentration of metabolites or in any of those physical characteristics which
are observable by ^{31}P NMR.

Short contractions: Recovery Fig. 5 shows the results of
experiments performed on toad gastrocnemii stimulated for
1 sec every 2 min. Spectrum A was obtained during the
first 16 sec following stimulation and spectrum B represents the last 32 sec of recovery, immediately before the
next stimulation. In order to reveal small differences

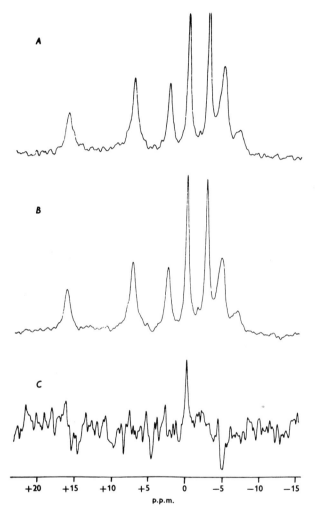

Fig. 5. Brief repeated contractions of toad muscle (1 sec/124 sec).
These spectra were accumulated in 5 separate experiments, each on a
pair of gastrocnemii. There was a total of 244 contractions; scans
every 2 sec. (A) First 16 sec following relaxation; 244 × 8 = 1952
scans. (B) Last 32 sec of recovery, before the next contraction; 244
× 16 = 3904 scans. The ordinate is divided by 2 for direct comparison
with A. (C) Spectrum A subtracted from spectrum B in the computer;
the ordinate is magnified 4-fold compared with A and B.

between the two spectra, record A was subtracted from
record B in the computer. The difference, scaled up 4-fold,
is shown in Fig. 5C. The most outstanding features of this

difference spectrum are the two peaks at 0 and -5 ppm,
showing the breakdown and subsequent rebuilding of about
10 per cent of the steady-state PCr content, corresponding
roughly (see Discussion) to a breakdown of 1.5 mmol kg^{-1}.
It is interesting that the chemical shift of the negative
Pi peak in the difference spectrum corresponds to a pH of
7.3, whereas the average pH as obtained from the Pi resonance in Figs. 5A and B is 7.1. The other feature of interest in the difference spectrum is the possible negative
peak in the region of the β ATP resonance.

The spectra shown in Fig. 5A and B differ from those
obtained on resting muscles in concentrations of PCr, Pi
and sugar P. This is because the gastrocnemii are too
thick to remain fully oxygenated during activity. When
well oxygenated sartorius muscles are similarly stimulated,
the spectra obtained differ only slightly from those of
resting muscles (see [7], Fig. 8); in particular, there
is no maintained increase in sugar P. In one experiment a
set of 4 sartorius muscles were subjected to 1 sec stimulations 422 times within the spectrometer. By the end of
this period the tension development fell to one quarter
of its original value and the concentration of Pi rose to
7 mmol 1^{-1}. There was still little increase in sugar P
above resting levels.

'Recovery' after metabolic poisoning

We are presently engaged in a series of experiments in
which some of the metabolic recovery processes are inhibited in order to determine the time course and extent of
those remaining. Glycolysis can be inhibited by addition
of 1 mM iodoacetic acid to the Ringer's solution; this drug
binds to the enzyme glyceraldehyde 3P dehydrogenase, causing a block which leads to the build-up of hexose phosphates. Oxidative reactions are blocked by bubbling the
Ringer's solution with nitrogen instead of oxygen. Because
it is very difficult to ensure that all of the oxygen is

removed we have also poisoned the muscles with 2 mM sodium cyanide, which blocks the utilisation of oxygen by the mitochondria.

We shall describe two types of experiments in which these inhibitors are used: (a) experiments in which both glycolytic and oxidative recovery processes are blocked, and (b) experiments in which only the oxidative reactions are blocked.

'Recovery' in iodoacetate, nitrogen and sodium cyanide

Fig. 6 shows a resting spectrum accumulated before, and three successive spectra accumulated after, a 15 sec contraction in frog gastrocnemius muscles. Unlike the case of well oxygenated, unpoisoned muscles, the PCr broken down as a result of contraction is not replenished in the post contractile period. The level of Pi decreases and that of sugar P increases during the 'recovery' period. These spectra also hint at an increase in the unknown compounds during the period 1–3 min following contraction (compare spectra A and C).

Fig. 7 shows that approximately half of the initial PCr is broken down as a result of contraction, and that its level then remains remarkably constant. The increase in Pi as a result of contraction equals the breakdown of PCr, as has been observed in unpoisoned muscles. However, in these experiments, unlike those on recovery of normal muscles, the loss of Pi during the 'recovery' period is mirrored, both in time-course and extent by an increase in sugar P. The total phosphorus distributed among the various metabolites remains constant within experimental error.

'Recovery' in muscles poisoned with NaCN In cyanide alone partial recovery of the PCr level does occur following contraction, approximately half the PCr broken down being restored. This is not surprising since the glycolytic pathway is intact and ADP can be rephosphorylated with the subsequent production of lactic acid and, to a lesser degree, of pyruvic acid. We cannot, of course, observe the

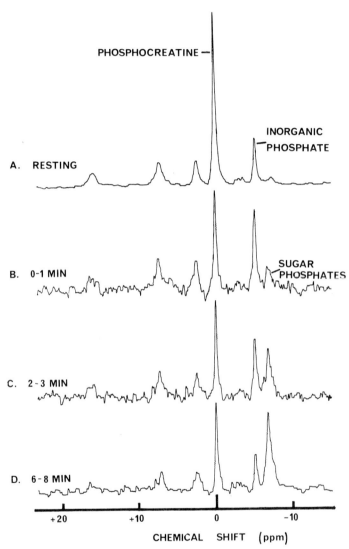

Fig. 6. 'Recovery' following a single 15 sec tetanus in poisoned frog gastrocnemius muscles. These spectra represent the sum of 4 experiments on 4 different pairs of frog gastrocnemii which had been poisoned for two hours with 1 mM iodoacetic acid and 2 mM sodium cyanide. Muscles were maintained at 4°C throughout the whole experiment. Changes were followed for 20 mins following contraction, and representative spectra from the first 8 mins are shown. The vertical scales are calculated in relation to the number of scans so that the spectra are directly comparable.

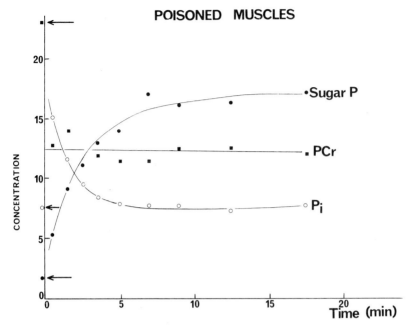

Fig. 7. Time-course of metabolic changes following contraction in poisoned muscles. The data presented are from the same experiments as shown in Fig. 6. The approximate concentrations shown on the ordinate are calculated on the reasonable assumption that the initial PCr concentration was 23 mmol kg^{-1}. Note the arrows indicating concentrations before stimulation.

production of these acids directly by ^{31}P NMR, but we can measure the resultant decrease in internal pH, approximately 0.15 in these experiments.

Discussion

In order to understand the initiation and control of cellular processes, it is necessary to know how these processes occur in intact cells under physiological conditions. NMR is being increasingly exploited for this purpose, as is indicated by the presentation of several papers on this topic at the present Conference. One of the most interesting new developments is that ^{31}P NMR is now being used to study tissues which perform their natural functions and can be subjected to various types of experi-

mental intervention within the spectrometer. This new area
of research is a logical consequence of the development of
methods to maintain tissues in physiological condition and
the ability to improve the time resolution by averaging
signals in synchrony with the experimental intervention.
The results presented in this paper and in this volume
illustrated only a small fraction of the experimental possibilities that have been created.

Comparison of NMR spectra with chemical analysis of resting muscle

The nature of the information yielded by NMR differs
slightly from that obtained by conventional chemical analysis. *Firstly*, NMR spectra can be obtained on intact living tissue, while chemical analysis requires prior freezing and extraction. Thus, conventional methods may be
subject to artifacts due to incomplete extraction or to
accidental breakdown of unstable metabolites. On the other
hand, quantitative interpretation of NMR spectra is subject to its own ambiguities, such as possible overlap or
saturation of resonances. *Secondly*, while total extract-
able metabolite is measured by chemical methods, the NMR
resonance linewidth is dependent upon molecular mobility,
so that highly immobilized compounds give rise to resonances which may be too broad to observe. In our spectra
the linewidths suggest that the resonances arise almost
exclusively from compounds which are mobile, and are
therefore probably in free solution. Thus, the essential
agreement between our results obtained by ^{31}P NMR and by
chemical analysis gives us much confidence that these
values do, in fact, represent the total concentration of
mobile metabolite in intact tissue.

pH Measurement by NMR

The two most commonly used techniques for the measurement of intracellular pH are 1) insertion of a pH sensitive

microelectrode directly into the cell, or 2) analysis of
the distribution of weak acids, from which pH can be calculated. Both of these methods have well documented dis-
advantages (see [23,24]). ^{31}P NMR provides an alternative
method, which is more direct than either of these, and has
the additional advantage that it yields information concerning the distribution of pH within the muscle.

Resting muscle The internal pH of the resting frog sartorius muscle under the conditions of our experiments was
determined from the position of the Pi resonance peak to
be 7.5. Muscle pH is dependent upon a number of factors,
including the pCO_2 and the pH of the external medium as
well as upon mechanical activity and metabolic state.
Tris-buffered Ringer's solution was used in these experiments, at a pH of 7.7; the pCO_2 was negligible. Our ^{31}P
NMR results agree reasonably well with those obtained in
other muscles under conditions similar to ours, using
microelectrodes [25] or the distribution of DMO (5.5
dimethyl-2, 4-oxazolidinedione) [26]. Due to the high external pH and lack of CO_2 in our experiments, it is very
likely that the pH of these muscles was somewhat greater
than would be found *in vivo*.

Contraction and recovery The measurement of muscle pH
during contraction and recovery can provide useful general
information concerning the rates and extents of metabolic
and mechanochemical processes. Dubuisson [27] and Distèche
[28] have measured changes in pH by placing a glass electrode on the surface of active muscle. Although variations
of approximately 0.001 pH unit can be detected by this
method, it is not clear exactly how these variations relate to the actual intracellular pH changes. In the present study, direct evidence has been obtained of variations
in intracellular pH following long and short contractions
in normal muscles and muscles poisoned with cyanide and
nitrogen.

In studies of repeated 1 sec contractions of toad

muscle spectra obtained immediately before and after contraction both indicate the same average intracellular pH
of 7.1 (see Fig. 5). The difference from the resting value
of 7.4 presumably results from the accumulation of lactic
acid. However, in the difference spectrum (Fig. 5C) the
peak corresponding to Pi is shifted by 0.2 pH unit to a
slightly more alkaline position. This may indicate that
there is, in fact, a shift immediately after contraction
but that it is too small to alter the observed average pH;
alternatively, it may possibly indicate that the Pi liberated during contraction is in a more alkaline environment
than the average for the muscle.

The pH changes observed after 25 sec contractions of
frog sartorii are rather ill-defined. This is because in
the first half of the recovery period the inorganic phosphate resonance is fairly broad, while in the latter half
it is very small. However, the average pH seems to fall to
about 7.0 during the first 20 minutes and then return to
its initial value of 7.5 after about 40 minutes. More interesting and better defined pH changes are observed in
muscles poisoned with cyanide and nitrogen, and these are
discussed below under the appropriate heading.

The Unidentified Resonances

In this and in previous ^{31}P NMR studies of skeletal
muscle [2,3,4,7] unexpected resonances have been observed
in the region between the PCr and Pi peaks, indicating the
presence of several previously unsuspected compounds
occurring in large quantities ($0.5-15$ mmol 1^{-1}). At least
four such unidentified resonances can be distinguished by
their different chemical shifts. Resonances at -2.7, -3.05,
and -3.6 ppm, in the phosphodiester region of the spectrum
occur to varying extents in amphibian, mammalian and avian
muscle [3,4,7]. The resonance at -3.05 ppm has been assigned to glycerophosphorylcholine [29,4]. In the present
study an additional small resonance has been observed at

-1.55 ppm in the spectra of toad gastrocnemii; compounds such as phosphoenolypyruvate and 1,3-diphosphoglycerate resonate in the region of -1.5 to -2.0 ppm at neutral pH.

The observation of these resonances is consistent with earlier detection in muscle of phosphorus compounds of unknown functional significance. In pioneering studies using paper chromatography, unexpected compounds were reported in muscles from frog [30], turtle [30,31], and human [32]. One such compound was identified in turtle muscle as the O-phosphodiester of L-serine and ethanolamine [31]; the others remain unidentified. The fraction of the total phosphorus found in these compounds agreed roughly with what is now indicated by the unidentified resonances in ^{31}P NMR spectra. It is thus likely that at least some of the same compounds are being studied.

In the present study we have focussed particularly on the question whether the unknown resonances change as a result of muscle contraction. In many experiments, small changes in these resonances seem to take place following contraction. For example, in the experiment shown in Fig. 6, the unknown resonances are more clearly defined in the intervals 1-2 min and 2-3 min than at other times. Unfortunately, the changes are not reproducible, and therefore it has proved difficult to verify their existence simply by adding together the result of a large number of experiments. A problem frequently encountered in NMR is to find an objective (rather than subjective) method of assessing the significance of a change in a resonance (whether it be in area), chemical shift or linewidth). We are now engaged in a statistical analysis of our results to establish with what certainty we can state that any apparent changes exist. If any changes in the unknown resonances are verified, then clearly these represent reactions occurring in muscle about which nothing is at present known.

NMR Studies of Stimulated Muscle

Recovery from 25 sec contractions Approximately 20 per
cent of the PCr (roughly 5 mmol l^{-1}) is broken down as a
result of a 25 sec contraction, and this is approximately
equalled by the increase in concentration of Pi. This
result agrees with the observation that about 1 per cent
of PCr is broken down per second of contraction [8]. Both
the concentrations of PCr and Pi recover with a half-time
of approximately 10 min. The same value for the half-time
of recovery of PCr was reported by Dydyńska and Wilkie
[33] when studying 30 sec contractions of the same muscle
under similar conditions. Kushmerick and Paul [34] also
found a half-time of about 10 min for recovery of PCr and
Pi as well as for recovery oxygen consumption in frog sartorii undergoing 25 sec contractions at 0°C.

Formation of Sugar Phosphates An unexpected finding is
the maintenance of sugar phosphates at a high and constant
level throughout the recovery period following long contractions. High levels of sugar phosphates are not observed in resting muscles, and indeed such an event is generally associated with anoxia and muscle deterioration. For
this reason, it must be emphasised that, in our experiments, the muscles are not depleted of ATP and PCr during
the latter half of the recovery period (see Fig. 4),
nor is their mechanical response declining.

In frog gastrocnemii most of the sugar phosphate is
generated after the first of a series of contractions.
This is consistent with the increasing evidence, obtained
by other methods, that first contractions differ from those
following. The recovery processes also differ according to
the pattern of stimulation (perhaps as a result of differences in phosphorylase activity [7]), for during a series of
1 sec contractions of frog sartorii, very little sugar
phosphate is generated. Chemical analysis, because it
relies on the difference between control and experimental
muscles, is most amenable to the study of single contrac-

tions. Since ^{31}P NMR can be used to study repeated contractions and the effects of different patterns of stimulation, the results obtained by this technique will help
to elucidate to what extent the information available
about single contractions can be applied generally.
Recovery from 1 sec contractions every 2 min Experiments
with the toad gastrocnemius permitted observation of clear
and consistent changes between these brief contractions.
Fig. 5C shows a rebuilding of about 10 per cent of the
steady-state concentration of PCr between 10 and 108 sec
following stimulation. Taking the composition of resting
toad muscle from Table 1 of Burt *et al*. [3], and assuming
that the rebuilding of PCr occurs at a steady rate throughout the 124 sec cycle, this corresponds to about 1.5 mmol
kg^{-1} PCr breakdown during each contraction. No measurements have been reported of the chemical changes during
contraction of toad gastrocnemius at 4°C, due to the technical difficulties of this type of experiment by chemical
methods. However, the value obtained by NMR (although it
requires further verification) indicates that in a steady
state of contraction at 4°C, the toad gastrocnemius breaks
down some two-and-a-half to five times more PCr per second
than do frog sartorii in a single brief contraction at 0°C
[8,35,36,19]. As indicated in the Introduction the breakdown of PCr in brief single contractions of frog and mammalian muscle is insufficient to account for the energy
produced and the deficit must be attributed to an unknown
process. The present results raise the possibility that
during a steady state of repeated contractions this deficit might disappear.

'Recovery' of Poisoned Muscles

Inhibition of both glycolytic and oxidative processes
These experiments were designed to investigate a 40 year
old problem in muscle physiology. In 1934 Lundsgaard (see
[37], p. 315) showed that in frog gastrocnemius muscles

poisoned with iodoacetate and nitrogen at 2°C about 30 per cent of the PCr splitting that resulted from a 25 sec tetanus occurred during the 3 min after contraction had ended. Since that time a number of studies have been designed to study this post-contractile PCr splitting, which has sometimes occurred and sometimes has not (see [37] pp. 225–6, and [38] pp. 331–2). Since the time-course of reactions can be followed in a single experiment using ^{31}P NMR, we have been able to obtain more complete results than have previously been available. Fig. 7 of this paper shows that there is no post-contractile PCr splitting in frog gastrocnemius muscles under the conditions of our experiments. We will not feel, however, that these results are conclusive until we have completed our efforts to confirm that the metabolic poisons were indeed 100 per cent effective. We are also reluctant to extrapolate these results to other types of muscle or to other conditions.

An interesting result of these experiments is that the rise in sugar P observed after contraction is accompanied by an equal loss of Pi, and no loss of PCr. It is not possible in the spectra of intact muscle to determine which sugar phosphates contribute to the sugar P peak. However, pH titrations of extracts of similarly treated muscles by ^{31}P NMR indicate that approximately 75 per cent of this peak probably arises from fructose diphosphate. Since formation of fructose diphosphate requires hydrolysis of ATP, we would have expected its formation to be accompanied by a breakdown of PCr, as well as a loss of Pi. We are currently analysing the identical extracts by conventional enzymatic methods to provide independent identification of the sugar phosphates.

Inhibition of oxidative metabolism The capability provided by ^{31}P NMR, and by no other technique, to measure simultaneously the absolute substrate levels and the pH provides a promising method of studying how the creatine phosphotransferase (CPT) reaction operates *in vivo*. This

in turn can provide information about the relative effectiveness of oxidative phosphorylation and of glycolysis in
rephosphorylating ADP. The reasoning is as follows: in the
CPT equilibrium

$$K_{eq} = \frac{[ATP][Cr]}{[ADP][PCr][H^+]} \tag{1}$$

all the terms on the right hand side can be determined
save for [ADP] since it is the free cytoplasmic concentration that is involved and this is known to be very much
less than that found by chemical analysis. The value of
K_{eq} is not accurately known for the conditions of our
experiments but is of the order 2×10^9 M^{-1}. Assuming
this, or any similar value, it is then possible to calculate [ATP]/[ADP], which is a measure of the effectiveness
of any rephosphorylating mechanism. Our preliminary experiments on this topic suggest that in normal resting
muscle [ATP]/[ADP] is large (e.g. 300–400). Poisoning with
NaCN (and thus forcing the muscle to depend on glycolysis)
reduces this ratio substantially (e.g. to about 100).
However, the incomplete recovery of PCr that follows activity in muscles poisoned with NaCN is not due to any sud-
den deterioration in phosphorylating effectiveness consequent on stimulation. It seems to result entirely from the
acid production that is so striking a sequel of contraction in these muscles and can be accounted for quantitatively by inserting the measured values of [H^+] into Equation (1).

*The Present Role and Future Prospects of NMR in the Study
of Living Muscle*

The use of ^{31}P NMR in the study of intact tissue is new
and fashionable. In the end, however, it must be justified
by the quality of the information it yields. In the study
of living contracting muscle, our *in vivo* measurements
using ^{31}P NMR have provided a completely independent con-

firmation of the validity of the methods of quick freezing,
extraction and chemical analysis that have been pains-
takingly evolved over many years. Our studies of the changes occurring during and following contraction have also
been useful for what they have shown does not occur. Speculation can be laid to rest concerning effects which we
would have seen, had they existed. Our current study of
the reactions following contraction in metabolically poisoned muscles are yielding information which should increase understanding of the control of metabolic processes.

The greatest problems encountered in the present study
have arisen from the low sensitivity of NMR, and the consequent poor time resolution. In studies of living muscle
severe constraints are imposed on the sample size (see
Methods), resulting in even smaller signals than is usual
in NMR studies. However, it is very likely that the quality of the spectra will be improved by future technical
advances. For example, wide bore magnets will permit the
use of larger muscles, perfused through their own vascular
system. In addition the situation may well be improved by
application of statistical methods to detection of small
signals in noise. With improved signal detection, the time
resolution could be increased by an extension of the method
of accumulating spectra during preset time intervals phased
to repeated contractions. The ultimate limit to time resolution, about 20 ms, is set by the allowable linewidths of
the phosphorus resonances.

The usefulness of NMR and of chemical analysis are complementary. Small changes in concentration of known compounds, particularly over short time intervals, are more
amenable to examination by chemical analysis. However, it
is already clear that ^{31}P NMR can provide information of a
diverse nature which could not be obtained by other techniques. We are now successfully using ^{31}P NMR in conjunction with chemical analysis to investigate problems in
muscle metabolism which would be beyond the scope of either

method alone.

Acknowledgements

We wish to thank Dr. R.E. Richards and Dr. G.K. Radda
for their support. Dr. D. Hoult and Mr. D. Gower gave in-
valuable assistance with the experimental work. The Paul
Fund of the Royal Society, Imperial Chemical Industries,
the Muscular Dystrophy Association of America, the Wellcome Trust, the National Science Foundation of America,
the Medical Research Council and the Science Research
Council provided essential financial support.

References

1. Hoult, D.I., Busby, S.J.W., Gadian, D.G., Radda, G.K., Richards,
   R.E. and Seeley, P.J. (1974). *Nature, Lond*. **252**, 285–7.
2. Bárány, M., Bárány, K., Burt, C.T., Glonek, T. and Myers, T.C.
   (1975). *J. Supramol. Struc*. **3**, 125–40.
3. Burt, C.T., Glonek, T. and Bárány, M. (1976). *J. Biol. Chem*. **251**
   2584–91.
4. Seeley, P.J., Busby, S.J.W., Gadian, D.G., Radda, G.K. and Richar
   Richards, R.E. (1976). *Transactions of the Biochemistry Society*,
   4, 62–4.
5. Burt, C.T., Glonek, T. and Bárány, M. (1977). *Science* **195**, 145–9.
6. Dawson, J. Gadian, D.G. and Wilkie, D.R. (1976). *J. Physiol*. **258**,
   82–3P.
7. Dawson, M.J., Gadian, D.G. and Wilkie, D.R. (1977). *J. Physiol*.
   (in press).
8. Gilbert, C., Kretzschmar, K.M., Wilkie, D.R. and Woledge, R.C.
   (1971). *J. Physiol*. **218**, 163–93.
9. Gadian, D.G., Garlick, P.B., Radda, G.K., Seeley, P.J. and Sehr,
   P.A. (1977). *These proceedings*.
10. Busby, S.J.W., Gadian, D.G., Radda, G.K., Richards, R.E. and
    Seeley, P.J. (1977). Submitted to *Biochem. J*.
11. Kretzschmar, K.M. and Wilkie, D.R. (1969). *J. Physiol*. **202**, 66–7P.
12. Wilkie, D.R. (1976). *Muscle Studies in Biology No. 11*, 2nd edn.,
    Edward Arnold, London.
13. Dawson, J., Gower, D., Kretzschmar, M.K. and Wilkie, D.R. (1975).
    J. Physiol. (in press).
14. Gower, D. and Kretzschmar, K.M. (1976). *J. Physiol*. **258**, 659–71.
15. Curtin, N.A., Gilbert, C., Kretzschmar, K.M. and Wilkie, D.R.
    (1974). *J. Physiol*. **238**, 455–72.
16. Rall, J.A., Homsher, E., Wallner, A. and Mommaerts, W.F.H.M.
    (1976). *J. gen. Physiol*. **68**, 13–27.
17. Woledge, R.C. (1975). *Pestic Sci*. **6**, 305–10.
18. Wilkie, D.R. (1975). *Ciba Fdn. Symp*. **31**, 327–39.
19. Kushmerick, M.J. and Paul, R.J. (1976). *J. Physiol*. **254**, 711–27.
20. Kushmerick, M.J. (1977). *Curr. Top. Bioenerg*. (in press).

21. Hill, A.V. (1965). *Trails and Trials in Physiology*. Edward Arnold, London.
22. Hoult, D.I. and Richards, R.E. (1975). *Proc. R. Soc. A* **344**, 311–40.
23. Waddell, W.J. and Bates, R.G. (1969). *Physiol. Rev.* **49**, 285–329.
24. Cohen, R.D. and Iles, R.A. (1975). CRC *Critical Reviews in Clinical Laboratory Science* **6**, 101–43.
25. Aickin, C.C. and Thomas, R.C. (1976). *J. Physiol.* **260**, 25–6P.
26. Roos, A. (1975). *J. Physiol.* **249**, 1–25.
27. Dubuisson, M. (1939). *J. -Physiol.* **94**, 461–82.
28. Distèche, A. (1960). *Mém. Acad. r. Belg. Cl. Sci. 8°*, série 2, tome, 32, pp. 1–169.
29. Burt, C.T., Glonek, T. and Bárány, M. (1976). *Biochemistry, N.Y.* **15**, 4850–3.
30. Caldwell, P.C. (1953). *Biochem. J.* **55**, 458–67.
31. Roberts, F. and Lowe, I.P. (1954). *J. Biol. Chem.* **211**, 1–12.
32. Caldwell, P.C. and Prankerd, T.A.J. (1954). *J. Neurol. Neurosurg. Psychiat.* **17**, 127–8.
33. Dydyńska, M. and Wilkie, D.R. (1966). *J. Physiol.* **184**, 751–69.
34. Kushmerick, M.J. and Paul, R.J. (1976). *J. Physiol.* **254**, 693–709.
35. Homsher, E., Rall, J.A., Wallner, A. and Ricchiuti, N.V. (1975). *J. Gen. Physiol.* **65**, 1–21.
36. Curtin, N.A. and Woledge, R.C. (1975). *J. Physiol.* **246**, 737–52.
37. Lundsgaard, E. (1934). *Biochem. Z.* **269**, 308–28.
38. Carlson, F.D., Hardy, D. and Wilkie, D.R. (1967). *J. Physiol.* **189**, 209–35.

SPATIALLY-RESOLVED STUDIES OF WHOLE TISSUES, ORGANS AND ORGANISMS BY NMR ZEUGMATOGRAPHY

PAUL C. LAUTERBUR

Department of Chemistry, State University of New York at Stony Brook, Stony Brook, N.Y. 11794, U.S.A.

Introduction

Living organisms are spatially complex at every level from the atomic scale to that of the whole organism, and the spatial complexity is coupled with chemical and physical heterogeneity. In the absence of techniques capable of resolving and following simultaneously all of the processes occurring in a functioning organism, a wide variety of methods has been developed for studying isolated chemical and functional subsystems. Although these methods have been remarkably successful, *in vivo* measurements of biochemical and biophysical changes should be much more efficient and less prone to error. In some applications, such as medical diagnosis, methods for non-invasive, non-interfering analysis and study are often required by the nature of the problem. Nuclear magnetic resonance spectroscopy is one of the most subtle and powerful methods for studying isolated biochemical and biophysical processes, because of its sensitivity to small changes in molecular structures, interactions and motions, and is beginning to be used for *in vivo* studies as well, where an additional advantage is the almost complete absence of the non-specific scattering and absorption that often impede the use of optical spectroscopic techniques in intact organisms. This latter advantage is a consequence, however, of the weakness of the interaction between the spins and the radiofrequency field, which can produce no appreciable amount of absorption in objects of reasonable size. This weak interaction is responsible for the relatively low sensitivity of NMR spectroscopy, which often severely limits its applicability to biological problems. An even more important limitation has been the lack of any method for spatially resolving the NMR signals, so as to distinguish among substances and properties in different regions of an organism. During the past several years, a number of methods for resolving local properties within NMR samples and for generating images from the resonance signals has been developed [1-26]. These methods all depend upon the encoding of spatial information into the NMR signals by the use of non-uniform magnetic fields, and are sometimes called zeugmatographic imaging to distinguish them from conventional imaging by optical methods [1]. It is probably possible to combine zeugmatographic imaging with all other NMR measurements, so that the specificity and subtlety of NMR spectroscopy can be brought to bear effectively on studies of intact and functioning organs and organisms, although the low sensitivity of NMR is even more of a limitation when the intensity in the usual 'zero-dimensional' spectrum must be spread over one, two or three spatial dimensions.

Techniques

Six general approaches to NMR imaging have been described. Techniques using reconstruction from projections [1,2,5,6,11,26] can be employed with either CW (continuous wave, or scanning) or transient (pulsed Fourier transform) data. In general, any technique that can give the NMR signal strength as a function of frequency in an inhomogeneous magnetic field may be used. NMR 'diffraction' [3,4,9] is a special technique for the study of periodic objects, such as crystals, and will not be discussed here.

Selective irradiation techniques [8,9,12,15,17.18,21-23, 25,26] use the time dependence of the selectively pertur- bed nuclear magnetization [27] to select only a portion of an object for observation. The sensitive point method (7,10,20] is capable of directly monitoring and scanning an object point-by-point in an appropriately modulated magnetic field. Fourier transform zeugmatography [13,14] makes use of the phase shifts produced by gradients ap- plied during the transient response of the spins, and 'FONAR' [19,24] achieves some spatial resolution by tech- niques that have not yet been fully described. More de- tailed descriptions of the different methods are given below.

Reconstruction from projections of the distribution of nuclear magnetization within an object, makes use of the fact that each point on a resonance line that has been broadened by an externally-applied magnetic field gradi- ent, represents all of the nuclei near a surface of con- stant field intersecting the sample. If the gradient is linear, the surfaces are equally-spaced planes, and the line shape is a simple one-dimensional projection, or double integration, of the three dimensional distribution of signal intensity. The resonance frequency or field then represents a spatial co-ordinate in the gradient direction. If the gradient vector is rotated about an axis, it gen- erates additional one-dimensional projections of the two- dimensional shadow of the three-dimensional object as if it were being viewed from along that axis, as shown in Fig. 1. Standard methods of image reconstruction from projections [28] may then be applied to invert the pro- cess, producing a two-dimensional image representing that view of the origins of the NMR signals within the sample. Reconstruction methods that have been used for NMR zeugma- tographic imaging include the convolution-filtered back projection method [29–32], and direct iterative algebraic (ART) methods [33–35]. In general, for a linear resolution

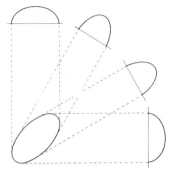

Fig. 1. One-dimensional projections of the two-dimensional shadow of a three-dimensional object. Each projection is the NMR signal distribution along the frequency axis in the presence of a linear magnetic field gradient. Reconstruction of the shadow may be accomplished by the various methods mentioned in the text.

of n elements across the two-dimensional image, approximately n equally-spaced projections should be used, although fewer may provide useful results in some special cases. A complete three-dimensional image may be reconstructed from a set of two-dimensional projections, as shown in Fig. 2. If each shadow is viewed from a different direction, but all of those directions lie in the same plane, a slice by slice reconstruction may be carried out with exactly the same algorithm as was used to generate

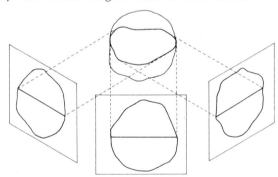

Fig. 2. Two-dimensional shadows of a three-dimensional object. Each shadow can be reconstructed from a set of one-dimensional projections generated by rotating a field gradient vector about the corresponding viewing axis. A slice by slice reconstruction, which may be accomplished by the usual methods, is the reverse of the process illustrated here, the edgewise projection of a slice onto a set of two-dimensional shadows.

each two-dimensional picture from its one-dimensional projections. Alternatively, the projection directions may be
distributed over all orientations, instead of being con-
fined to a plane. This more general approach to three-
dimensional reconstruaction may be useful or even necessary under some circumstances [36]. Of course, an image
with, for example, one per cent resolution in each dimension contains one million picture elements, and the problems with sensitivity, computations, and display are not
trivial.

One-dimensional views, or projections, may be generated
directly by scanning the spectrum, or may be derived from
a free induction decay or a spin echo by the usual Fourier
transform techniques. When the NMR signal is acquired
initially as a function of time, the free induction decay
or spin echo may be multiplied by the Fourier transform of
the convolution filter function before a back projection
is done to form the image [11,26]. Reconstruction methods
are relatively simple to implement, given modern minicom-
puters, and are quite versatile. They have the advantage
of acquiring data simultaneously from all parts of the
sample, although the sensitivity that results from this
'multiplex advantage' is decreased by the noise enhancement that accompanied reconstruction [32,37]. Combinations of reconstruction with selective excitation are also
possible [9,17,18].

Many selective excitation, or, more generally, selective irradiation, methods have been proposed. They make
use of the fact that a radiofrequency pulse of a given
shape effectively contains a range of frequencies inversely
proportional to its width. If a signal is spread over a
wide frequency range by a field gradient, a long pulse will
produce a perturbation largely restricted to the vicinity
of a single plane of constant field. If the first gradient
is then removed, and a second gradient applied perpendicular to the direction of the first, the signals, either

free induction decays or pseudo-echoes [27], within the
excited slice will be resolved along a second co-ordinate.
Two dimensional shadow images may be obtained directly from
such an experiment [8,12,15], or it may be combined with
reconstruction to give images of selected slices [8].
Special pulse sequences capable of directly generating
images very rapidly have been described [26,30], and may
be especially useful when the total measurement time is the
most significant factor in the experiments.

Although two linear static gradients merely combine vec-
torially to give a new linear static gradient, and therefore give only another one-dimensional projection, time-
dependent gradients can encode two- or three-dimensional
information into the signal in such a way that it can be
recovered without reconstruction calculations. A particularly elegant analogue technique for the partial recovery
of spatially-resolved signals has been called spin mapping
[7,10], or the sensitive point method [20]. It makes possible a point by point scan in any desired pattern through
a three-dimensional object, or the continuous monitoring
of the NMR signal intensity or relaxation times at a single point. Because only a small portion of the object is
contributing to the signal at any given instant, however,
the original single sensitive point method will usually be
relatively slow if complete two- or three-dimensional
images are required.

Another way of using time-dependent gradients has been
called Fourier transform zeugmatography [13,14], and may
be considered to be a form of multi-dimensional spectroscopy [38,39]. Preservation of the signal phase information, as two or three orthogonal gradients are applied during each free induction decay, permits the recovery of all
spatial information by a simple two- or three-dimensional
Fourier transformation. This approach to the problem may
have the greatest theoretical sensitivity of any possible
zeugmatographic imaging technique.

Limitations

For a given magnetic field gradient, the useful resolution is determined by T_2, the spin-spin relaxation time, and by magnetic field distortions, possibly as large as 10 ppm, caused by magnetic susceptibility differences. The resolution can be increased by using larger gradients, but at the expense of sensitivity. Regardless of the imaging technique used, the resolution, selectivity, and discrimination achievable will almost always be limited by the signal to noise ratio. The maximum sensitivity possible may be estimated by calculating, for a particular apparatus, the signal to noise ratio for a volume of material whose dimensions are given by the desired resolution. This estimate ignores the degradation in effective sensitivity that may result from the special experimental constraints or data processing required by the imaging technique used, but is useful in that it gives an upper limit for the sensitivity and resolution attainable with a particular apparatus, independent of the details of a particular experiment.

Very general considerations lead to the result that the fractional resolution achievable decreases as an object becomes smaller. The practical limit for water proton resonances may be reached in the vicinity of 10 micrometers, even assuming the use of superconducting magnets and miniature receiver coils. It is therefore unlikely that the resolution of the light microscope can be reached, although the predicted resolution of 0.01 mm could be useful in histological studies above the cellular level.

Studies of very large biological objects will eventually by limited by the penetration depth of the radiofrequency radiation [40]. For the whole human body, practical limits may be of the order of 10 MHz. At such frequencies, a resolution of a few millimeters may be achieved for signal to noise ratios in the range 10 to 100 (probably sufficient to resolve many anatomical structures and relaxation time differences) and total scanning time of 10 minutes or less.

For the head and limbs, for isolated organs, and for small
animal studies, the greater sensitivity potentially achievable at higher frequencies may be realizable, and would be
well worthwhile. For example, an increase from the 4 MHz
frequency now being used for large objects to 40 MHz would
give a theoretical increase of a factor of 56 in sensitivity [41], or a reduction by a factor of over 3000 in the
time required to achieve a given sensitivity.

The advantages of higher magnetic fields may be realized
more readily when nuclei other than protons (and fluorine)
are observed, because the corresponding frequencies are
lower. It is therefore appropriate to compare the theoretical sensitivities at a constant frequency [42], if estimates of the potential for the use of relatively dilute
nuclides are to be made. Fig. 3 shows an estimate [43] of
the spatial resolution that may be obtainable for some potentially useful isotopes of common elements. It has been
assumed that the magnetic field has been adjusted to give
a 4 MHz resonance frequency, that the receiver coil has a
diameter of 30 cm, and that the total signal averaging time

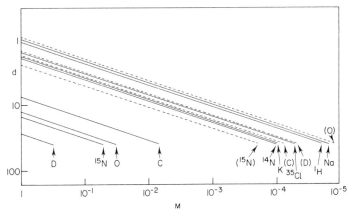

Fig. 3. Estimated relationships between the spatial resolution d(cm)
and the concentration M (moles/liter) for various nuclides. The calculations were for a resonance frequency of 4 MHz, a receiver coil diameter of 30 cm, and an averaging time of 1000 seconds, assuming that
$T_1/T_2 = 1$. The solid lines represent the isotopes in their natural
abundance, and the dotted lines those enriched to 100 per cent.

is 1000 sec. No effects of the relaxation times have been
included in the calculations, except for the assumption
that $T_1 = T_2$. It can be seen that physiologically realistic
concentrations of many elements may be detectable with
enough spatial resolution to selectively monitor some
organs or regions of the body. Higher frequencies, smaller
receiver coils, and longer averaging times, which may be
feasible in specialized biomedical studies, could provide
significantly better resolution. Nevertheless, it seems
likely that the proton resonances from water and fats will
be the most widely used in medical diagnostic applications
in the foreseeable future because their greater strength
allows much higher resolutions to be achieved in a given
time, and the morphological and water proton relaxation
time differences revealed by such images may be as useful
as much of the information that could be obtained at lower
resolution from other signals.

Examples

Practical apparatus for zeugmatographic imaging of
large objects is still under development. The pictures in
Fig. 4 illustrate the performance that has been achieved
with a partly-assembled system based on a special O.S.
Walker four-coil electromagnet, operating at 938 gauss,
with a 42 cm minimum bore. The objects used in these tests
were rotated in a constant gradient corresponding to 194
Hz/cm. The single radiofrequency coil was about 15 cm in
diameter, and a SEIMCO AF-74 pulsed NMR spectrometer was
used, operating at 4 MHz. Each of the 65 projections was
obtained by Fourier transformation of the average of 100
free induction decays following 20° pulses. The reconstruction was carried out on a DEC PDP 11/45 computer, using a
filtered back-projection method [44] on a 65 by 65 array,
linearly interpolated to form a 256 by 256 array, and displayed with the aid of a Ramtek digitally-refreshed video
colour display system. The resolution, approximately 2 mm

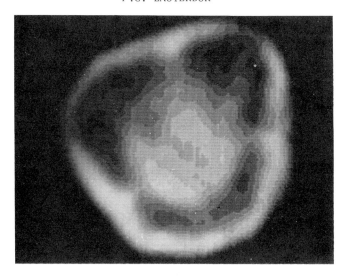

Fig. 4a

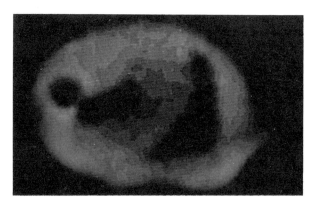

Fig. 4b

Fig. 4. 4 MHz proton NMR shadow zeugmatograms reconstructed from 65 projections, each one the Fourier transform of the average of 100 free induction decays, using a 15 cm diameter radiofrequency coil. (a) One-half of a green pepper. (b) A 1.8 cm thick slice cut from a calf's heart.

in the shadow plane, is probably limited by the coarseness of the original array and the limited number of projections used. Although several compromises were made in these pre-liminary experiments, it is evident that quite convincing two-dimensional images can be obtained.

Prospects

In the near future, images of the interiors of biological objects in terms of those properties that affect NMR signals will be generated routinely. Low-resolution zeugmatographic microscopy should be possible on small objects, and images for medical diagnostic imaging comparable to those recently achieved with X-rays by computed tomography are a distinct possibility. Even studies on minor chemical components of organs and organisms, and of elements other than hydrogen, may eventually be possible with special equipment. The detailed studies of chemical and physical processes within intact living organisms made possible by the addition of spatial resolution to conventional NMR spectroscopy may be expected to find a number of specific applications. Malignant tumours growing in mice have already been detected by virtue of their unusual water proton spin-lattice relaxation times [45] and the effect of a transplanted tumour on the signal intensity from the thoracic cavity of a mouse has been observed [19,24]. Preliminary experiments on the feasibility of studying lung edema by NMR zeugmatographic imaging are also in progress [46, 47].

Acknowledgements

This investigation was partially supported by Grant No. CA-153000, awarded by the National Cancer Institute, DHEW, and by Contract NO1-HV-5-2970, awarded by the National Heart, Lung and Blood Institute, DHEW. Acknowledgement is made to the Donors of The Petroleum Research Fund, administered by the American Chemical Society, for partial support of this research, and also to the Gulf Oil Foundation.

References

1. Lauterbur, P.C. (1973). *Nature* **242**, 190–1.
2. Lauterbur, P.C. (1973). *In: Proc. First International Conference on Stable Isotopes in Chemistry, Biology and Medicine*, AEC, CONF-730525, 255–60.

3. Mansfield, P., Grannell, P.K., Garroway, A.N. and Stalker, D.C. (1973). *In: Proc. First Specialized Colloque Ampere*, Institute of Nuclear Physics, Krakow, Poland, 16–27.

4. Mansfield, P. and Grannell, P.K. (1973). *J. Phys. C: Solid State Phys.* **6**, L422–6.

5. Lauterbur, P.C. (1974). *Pure Appl. Chem.* **40**, 149–57.

6. Lauterbur, P.C. (1974). *In: Techniques of Three-Dimensional Reconstruction: Proceedings of an International Workshop* (R.B. Marr, ed.), Brookhaven National Laboratory Publication BNL 20425, 20–22.

7. Hinshaw, W.S. (1974). *Phys. Letters*, **48A**, 87–8.

8. Lauterbur, P.C., Dulcey, C.S., Jr., Lai, C.-M., Feiler, M.A., House, W.V., Jr., Kramer, D., Chen, C.-N., and Dias, R. (1975). *In: Magnetic Resonance and Related Phenomena, Proceedings of the 18th Ampere Congress* (P.S. Allen, E.R. Andrew, and C.A. Bates, eds.), North-Holland, Amsterdam, Vol. 1, 27–9.

9. Mansfield, P., Grannell, P.K. and Maudsley, A.A. (1975). *In: Magnetic Resonance and Related Phenomena, Proceedings of the 18th Ampere Congress* (P.S. Allen, E.R. Andrew, and C.A. Bates, eds.), North-Holland, Amsterdam, Vol. 2, 431–2.

10. Hinshaw, W.S. (1975). *In: Magnetic Resonance and Related Phenomena, Proceedings of the 18th Ampere Congress* (P.S. Allen, E.R. Andrew, and C.A. Bates, eds.), North-Holland, Amsterdam, Vol. 2, 433–4.

11. Hutchison, J.M.S., Mallard, J.R. and Goll, C.C. (1975). *In: Magnetic Resonance and Related Phenomena, Proceedings of the 18th Ampere Congress* (P.S. Allen, E.R. Andrew, and C.A. Bates, eds.), North-Holland, Amsterdam, Vol. 2, 283–4.

12. Garroway, A.N., Grannel, P.K., and Mansfield, P. (1974). *J. Phys. C: Solid State Phys.* **7**, L457–62.

13. Kumar, A., Welti, D., and Ernst, R.R. (1975). *Naturwiss.* **62**, 34.

14. Kumar, A., Welti, D., and Ernst, R.R. (1975). *J. Mag. Res.* **18**, 69–83.

15. Lauterbur, P.C., House, W.V., Jr., Kramer, D.M., Chen, C.-N., Porretto, F.W., and Dulcey, C.S., Jr. (1975). *In: Image Processing for 2-D and 3-D Reconstruction from Projections: Theory and Practice in Medicine and the Physical Sciences*, Opt. Soc. Am. MA10-1 to MA10-3.

16. Lauterbur, P.C., Kramer, D.M., House, W.V., Jr., Chen, C.-N. (1975). *J. Am. Chem. Soc.* **97**, 6866–8.

17. Grannell, P.K. and Mansfield, P. (1975). *Phys. Med. Biol.* **20**, 477–82.

18. Mansfield, P., Maudsley, A.A., and Baines, T. *J. Phys. E: Sci. Instrum.* **9**, 271–8.

19. Damadian, R., Minkoff, L., Goldsmith, M., Stanford, M., and Koutcher, J. (1976). *Physiol. Chem. & Phys.* **8**, 61–5.

20. Hinshaw, W.S. (1976). *J. Appl. Phys.* **47**, 3709–21.

21. Mansfield, P. and Maudsley, A.A. (1976). *Phys. Med. Biol.* **21**, 847–52.

22. Baines, T. and Mansfield, P. (1976). *J. Phys. E: Sci. Instrum.* **9**, 809–11.

23. Mansfield, P. and Maudsley, A.A. (1976). *J. Phys. C: Solid State Phys.* **9**, L409–11.

24. Damadian, R., Minkoff, L., Goldsmith, M., Stanford, M., and Koutcher, J. (1976). *Science* **194**, 1430–2.

25. Mansfield, P. and Maudsley, A.A. (1977). *Brit. J. Radiology* **50**, 188—94.
26. Hutchison, J.M.S. (1977). *In: Medical Images: Formation, Perception and Measurement* (George A. Hay, ed.), Wiley, New York, 135—41.
27. But see Hoult, D.I. (1977), *J. Mag. Res.* **27**, 165—7 for a *caveat*.
28. For general reviews, see Brooks, R.A. and Di Chiro, G. (1976), *Phys. Med. Biol.* **21**, 689—732; *Radiology* **117**, 561—72 (1975), and Gordon, R., Herman, G.T. and Johnson, S.A. (1975), *Sci. Am.* **223**, (4), 56—68.
29. Bracewell, R.N. and Riddle, A.C. (1967). *Astrophys. J.* **150**, 427—34.
30. Ramachandran, G.N. and Lakshminarayanan, A.V. (1971). *Proc. Natl. Acad. Sci. USA* **68**, 2236—40.
31. Cho, Z.H., Ahn, I., Bohm, C. and Huth, G. (1974). *Phys. Med. Biol.* **19**, 511—22.
32. Shepp, L.A. and Logan, B.F. (1974). *IEEE Trans. Nucl. Sci.* **NS-21** (3), 21—43.
33. Gordon, R., Bender, R., and Herman, G.T. (1970). *J. Theor. Biol.* **29**, 471—81.
34. Gordon, R. and Herman, G.T. (1974). *Int. Rev. Cytol.* **38**, 111—51.
35. Gordon, R. (1974). *IEEE Trans. Nucl. Sci.* **NS-21** (3), 78—93.
36. For a brief review and references, see Radulovic, P.T. and Vest, C.M. (1975). *In: Image Processing for 2-D and 3-D Reconstruction from Projections: Theory and Practice in Medicine and the Physical Sciences*, Opt. Soc. Am. TuD1-1 to TuD1-4.
37. Tanaka, E. and Inuma, T.A. (1975). *Phys. Med. Biol.* **20**, 789—98.
38. Aue, W.P., Bartholdi, E., and Ernst, P.R. (1976). *J. Chem. Phys.* **64**, 2229—46.
39. Bodenhausen, G., Freeman, R., Niedermeyer, R., and Turner, D.L. (1977). *J. Mag. Res.* **26**, 133—64.
40. For reference and a general discussion, see Johnson, C.C. and Guy, A.W. (1972). *Proc. IEEE* **60**, 692—718.
41. Assuming a dependence on $w_0^7/4$: Hoult, D.I. and Richards, R.E. (1976). *J. Mag. Res.* **24**, 71—85.
42. Emsley, J.W., Feeney, J., and Sutcliffe, L.H. (1965). *High Resolution Nuclear Magnetic Resonance Spectroscopy*, Pergamon, Oxford. Vol. 1, Appendix A, 589—94.
43. Following Abragam, A. (1961). *The Principles of Nuclear Magnetism*, Clarendon Press, Oxford. 82—4.
44. Brooks, R.A. and Di Chiro, G. (1975). *Radiology* **117**, 561—72.
45. Lauterbur, P.C., Lai, C.-M., Frank, J.A., Dulcey, C.S., Jr. (1976). *Physics in Canada* **32**, 33.11.
46. Lauterbur, P.C., Frank, J.A. and Jacobson, M.J. (1976). *Physics in Canada* **32**, 33.9.
47. Frank, J.A., Feiler, M.A., House, W.V., Lauterbur, P.C., and Jacobson, M.J. (1976). *Clinical Research* **24**, 217A.

ABSTRACTS

1. NMR STUDIES OF RABBIT SKELETAL TROPOMYOSIN

B.F.P. EDWARDS and B.D. SYKES

*MRC Group on Protein Structure and Function,
Department of Biochemistry, University of Alberta,
Edmonton, Alberta, T6G 2H7, Canada*

The structure of αα-tropomyosin from rabbit muscle is
specified by its known sequence of 284 amino acids (Stone,
D. *et al.*, in *Proc. IX FEBS Meeting*, 1975, Proteins of Contractile Systems) and its probable coiled-coil structure
(Crick, F.H.C., *Acta Cryst.* **6**, 689, 1953). We have studied
the renatured dimer using proton nuclear magnetic resonance
at 270 MHz and have observed, through linewidths and relaxation times, a high degree of segmental motion in this
rod-like molecule. The symmetric histidine resonances in
our spectra support the finding that the two chains are in
register. Since each chain of the dimer has one histidine
residue near the centre (His 153) and one near the C-
terminal end (His 276), we have assigned the resonances
using carboxypeptidase digestion. The pair of central histidines titrate in a cooperative manner with Hill coefficients approaching 2.0 under certain conditions. The resonances of the two His 276 residues, which are thought to
be in the overlap region of polymerized tropomyosin, are
perturbed by polymerization at low ionic strength.

2. PMR STUDIES OF RABBIT SKELETAL TROPONIN:
PROTEIN—PROTEIN INTERACTIONS

B.A. LEVINE,* D. MERCOLA,** J.M. THORNTON[†]

**Inorganic Chemistry Laboratory, Oxford University
**ARC Unit, Zoology Department, Oxford University
[†]Laboratory of Molecular Biophysics, Zoology Department,
Oxford University*

Contractile response to Ca(II) in skeletal muscle is regu-
lated by the troponin-tropomyosin protein complex on the
thin filaments (Weber and Murray, *Physiol. Rev.* **53**, 612,
1973). Ca(II) binding to the TN-C component of troponin
triggers a conformational change which is transmitted
through the TN-I and TN-T subunits of the troponin trimer
complex.
   Ca(II) titration of the binding sites of TN-C correlated with triggering ATPase activity showed little altera-

tion of the internal protein conformation. This observation suggests that binding at these sites modifies the
orientation of surface groups and that these sites may well
be located close to TN-I and TN-T. The conformation of
TN-C was altered to different extents by TN-I and TN-T, the
spectral changes being greater in the presence of Ca(II).

A conformational change in TN-I was observed upon Ca(II)
binding to TN-C, indicating that the Ca(II) induced changes
in TN-C are transmitted to the adjacent TN-I. The reduction in relaxation times of arginine and lysine resonances
of TN-I and TN-T demonstrate interaction of basic regions
of these proteins with acidic sites on TN-C. The PMR spectrum of the troponin trimer (molec. wt. 80,000) is well resolved. The C2 and C4 resonances of 8 of the 11 histidine
residues are readily observed. These are separately titratable with pK_a's in the range 7.6-8-8. Several phenylalanine
and tyrosine resonances have also been identified.

3. COMPARISON OF THE COMBINING SITES OF
SEVERAL IMMUNOGLOBULINS

RAYMOND A. DWEK, PETER GETTINS, ARABELLA MORRIS,
SIMON WAIN-HOBSON, CAROLYN WRIGHT, and DAVID GIVOL*

*Department of Biochemistry, University of Oxford
*The Department of Chemical Immunology,
The Weizmann Institute of Science, Rehovot, Israel*

The crystal structure of the immunoglobulins McPC 603,
New and others have established the common structural
feature of the immunoglobulin fold. Similarity of the sequence of MOPC 315, a dinitrophenyl binding mouse myeloma
protein, to McPC 603 and New has enabled a model of the
combining site of MOPC 315 to be constructed. Perturbation
of aromatic residues and large shifts on the Dnp ring observed by NMR has confirmed the highly aromatic combining
site of the model.

Two other Dnp binding mouse myeloma proteins, MOPC 460
and XRPC 25 have also been shown to have similarly aromatic
combining sites. In contrast McPC 603, a phosphorylcholine
binding mouse myeloma, shows relatively few aromatic residues perturbed by hapten binding. In no case is there evidence for a major conformational change on hapten binding.

4. STRUCTURE OF METALLOTHIONEINS STUDIED BY CIRCULAR DICHROISM, 270 MHz PMR- AND EPR-SPECTROSCOPY

HEINZ RUPP, RICHARD CAMMACK and ULTICH WESER*

King's College London, School of Biological Sciences,
68 Half Moon Lane, London S.E.24 9JF, England
*Anorganische Biochemie, Physiologisch-chemisches Institut
der Universität, D-7400 Tübingen, Hoppe-Seyler-Strasse 1, GFR

Metallothionein belongs to a group of low-molecular-weight metalloproteins, which are characterized by a high cysteine content (20–30 per cent) and the binding of Zn, Cd and Cu. Cu-thionein from yeast exists exclusively as the copper protein, whereas the thioneins from animal tissues bind Zn, Cd and Cu depending on the nutritional status. The structural properties of Cu-thionein and Cd,Zn-thionein are compared using circular dichroism and 270 MHz PMR-spectroscopy. CD-measurements show that native Cd, Zn-thionein has a random-coil conformation. Deprotonation of the lysine-residues (12 per cent) of the apoprotein, cysteine-thionein, does not result in a change of the tertiary structure. In contrast, Cu-thionein from yeast exists to a high portion in the pleated sheet form. Upon treatment with chaotropic agents an unordered structure is seen. This pleated sheet structure can be considered the possible cause of the homogeneity with regard to the metal content of the microbial Cu-thioneins.

The chiroptical data are correlated with 270 MHz PMR. The actual spectrum of native Cd,Zn-thionein and the calculated spectrum of a hypothetical random coil thionein agree quite well. The reversible replacement of Cd and Zn by H^+ proved convenient for the assignment of proton signals, which are influenced by the metal binding (1.52 ppm, 1.03 ppm, 0.87 ppm). The cysteine-resonance of 2.95 ppm is shifted by 0.07 ppm to lower field upon binding of Cd. This downfield shift induced by the binding of diamagnetic metal ions was corroborated by ^{13}C-NMR-spectra of Cd-penicillamine chelates. The marked broadening of the 3.02 ppm signal suggests a rigid structure of the metal clusters. The relationship between metallothionein and the iron-sulphur proteins, especially rubredoxin, was studied by EPR-spectroscopy. Fe was employed for probing the metal-binding site in metallothionein. Zn of native Cd,Zn-thionein was successfully replaced by Fe. The Fe,Cd-thionein shows a signal at g = 4.3, which is characteristic for high-spin Fe(III) in a rhombic symmetry. Models for the metal clusters in metallothionein are discussed in relation to the iron-sulphur proteins.

This study was supported by a NATO research grant (No. 1256). H.R. was awarded a Research Fellowship (Ru 245/1) by the Deutsche Forschungsgemeinschaft.

5. ELECTRON TRANSFER CHAIN STUDIES OF SULPHATE REDUCING BACTERIA

J.J.G. MOURA and A.V. XAVIER

Centro de Química Estrutural da Universidade de Lisboa, I.S.T., Lisboa 1, Portugal

The electron transfer chain from H_2 to sulphite can use the following pathway:

$$H_2 \rightarrow \text{Hydrogenase} \rightarrow \text{Cytochrome } c_3 \rightarrow \text{Ferredoxin} \rightarrow SO_3^=$$

Three oligomeric forms of *Desulphovibrio gigas* ferredoxin have been isolated. They are two different trimers (FdI and FdI') and one tetramer (FdII) of the same basic unit of 6,000 molecular weight. High resolution NMR, magnetic susceptibilities and EPR studies show that the three forms can use different oxidation states (C^-, C^{2-}, C^{3-}) having always negative redox potentials. FdII stabilizes almost exclusively the C^-/C^{2-} states ($E_0 \simeq -50$ mV), FdI the C^{2-}/C^{3-} states ($E_0 \simeq -430$ mV), and FdI' has an intermediate behaviour.

Biological activity measurements proved that the tetramer (FdII) is more efficient in the sulphite reduction. Cytochrome c_3 has also four redox centres (haems) and uses a redox potential more positive than that normally used by bacterial ferredoxins ($E_{0(C^{2-}/C^{3-})} \simeq -400$ mV).

This suggests that *D. gigas* Fd has to tetramarize in order to obtain the necessary number of electrons at the necessary redox potential to accept the electrons from cytochrome c_3. To study the mechanism of electron transfer we have been investigating the interaction between cytochrome c_3 and FdII by high resolution NMR (in collaboration with Professor R.J.P. Williams and Drs. D.J. Cookson and G.R. Moore). Thus selective shifts of the contact shifted resonances of cytochrome c_3 were observed upon titration with FdII. Also the pattern of reoxidation of cytochrome c_3 shows important differences when it is observed in the presence and in the absence of FdII.

6. NUCLEAR MAGNETIC RESONANCE STUDY ON FLAVODOXINS

C.G. VAN SCHAGEN

Department of Biochemistry, Agricultural University, Wageningen, The Netherlands

Flavodoxin from the bacteria Peptostreptococcus elsdenii, which contains one molecule of riboflavin-5'-monophosphate (FMN) as a prosthetic group and possesses a molecular weight of 15000, has been investigated by NMR. The proton spectrum of the holoenzyme differs drastically from that of the apoenzyme indicating that a large conformational

change of the protein occurs upon removal of FMN. Furthermore, in the spectrum of the holoenzyme four well resolved
resonances are observed between zero and -1 ppm after convolution difference. With the aid of difference spectra,
where flavosemiquinone served as a paramagnetic centre,
and various pulse sequences two of these resonances could
be assigned to an isoleucine and an alanine. In the same
way the resonances due to the two methylgroups of FMN,
those of tyrosine and that of N(3)H were assigned. Furthermore ^{13}C-NMR measurements have been conducted employing
selectively ^{13}C enriched FMN derivatives.

The phosphate group of FMN served as a conformational
probe of the protein. The ^{31}P-NMR data indicate a strong
interaction between prosthetic group and apoenzyme. From
the measured T_1 value a τ_c of 10 nsec is computed which is
in agreement with the expected value for a protein of this
size.

From these data several conclusions could be reached
concerning the structure of the protein (^{1}H, ^{31}P), the
conformation and electronic structure of FMN (^{13}C), and
the active centre of the protein (^{1}H).

7. THE STRUCTURE OF CYTOCHROME c: A COMPARATIVE NMR STUDY

R.P. AMBLER,* J. CHIEN,[†] D.J. COOKSON,[‡] L.C. DICKINSON,[†]
G.R. MOORE,[‡] R.C. PITT[‡] and R.J.P. WILLIAMS[‡]

*Department of Molecular Biology, University of Edinburgh
[†]Department of Chemistry, University of Massachusetts
[‡]Inorganic Chemistry Laboratory, University of Oxford

The structure of cytochrome c in solution has been investigated by NMR spectroscopy. A major problem with NMR is the
specific assignment of resonances and this problem has
been illustrated for horse cytochrome c by assignment of a
selection of resonances. The resonances of trp 59 and phe
82 were chosen because these residues had been implicated,
from early X-ray crystallographic studies, to be mechanis-
tically important. In addition, the resonances of tyr 48
and phe 46 were chosen because their unusual properties
allow dynamic structural information to be obtained.

A further complication in comparison of the NMR spectra
of ferricytochrome c and ferrocytochrome c is the paramagnetism of the oxidized form. One approach to this problem is the study of metal substituted cytochromes c. We
have studied the diamagnetic protein, horse cobalticyto-
chrome c and comparison of the spectra of horse ferri-,
ferro- and cobalticytochromes c show that there is little
change in the internal conformation of the protein resulting from a change in oxidation state. However, the physical and ion-binding properties of horse cytochrome c suggest that there are differences in the conformation of the
surface between oxidized and reduced forms.

This study has been extended to bacterial cytochrome c: *P. aeruginosa* c_{551}, *P. mendocina* c_{551}, *P. stutzeri* c_{551}, *P. fluorescens* c_{551}, *Halotolerant micrococcus* c_{554}, *R. rubrum* c_2, *R. molischianum* c_2 (iso-1), *R. viridis* c_2 and *R. fulvum* c_2 (iso-1).

The results show that the structures of the proteins about the heme groups are remarkably similar.

8. THE ^{1}H NMR SPECTROSCOPY OF SUPEROXIDE DISMUTASE

A.E. CASS, H.A.O. HILL, B.E. SMITH,
W.J. BANNISTER and J.V. BANNISTER

*Inorganic Chemistry Laboratory, Oxford and
Department of Physiology and Biochemistry, Malta*

The class of metalloenzymes known as superoxide dismutases are believed to be responsible for the *in vivo* scavenging of the superoxide radical anion:

$$2O_2^- + 2H^+ \ \cdots\cdots\rightarrow\ O_2 + H_2O_2$$

The enzyme from the cytosol of eukaryotes is a protein composed of two identical subunits of total molecular weight 31,200, containing one copper and one zinc/subunit. We have studied bovine superoxide dismutase using 270 MHz NMR, both as a 'model' system for investigating the NMR spectra of other copper proteins and for probing structure-function correlations in enzymes.

A comparison of apo-, copper (I)-zinc(II)- and copper (II)-zinc(II)- forms of the protein has enabled us to propose assignments to ten amino acid residues. The data require that at least 4 and probably 6 histidine residues serve as ligands to the metals in each subunit of the enzyme, consistent with X-ray diffraction results. The remaining assignments are associated with His 19, His 41, Tyr 108 and the N-acetyl group of the N-terminal alanine.

One of the imidazole C2 protons exchanges with the solvent readily at pH*>8, and this has allowed the use of tritium exchange, coupled with proteolytic digestion and peptide mapping, in the assignment of this resonance to His 41, independently of knowledge of the X-ray diffraction results.

The structural implication of the effect of the paramagnetic copper(II) in the holoenzyme on the proton relaxation times are in qualitative agreement with the metal-proton distances as determined by X-ray diffraction studies.

9. ^{13}C NMR STUDIES OF VIRUS PROTEINS

J.L. de WIT, N.C.M. ALMA, M.A. HEMMINGA and T.J. SCHAAFSMA

*Department of Molecular Physics, Agricultural University,
De Dreijen 6, Wageningen, The Netherlands*

The usefulness of natural abundance ^{13}C NMR on proteins
with M ~ 50000 has been demonstrated by Oldfield *et al.*
[1]. For proteins that occur in assemblies, such as coat
proteins of viruses, natural abundance ^{13}C NMR spectra
will be more difficult to measure, even at high magnetic
field (i.e. 80 KG), because of the increase of linebroadening and T_1. ^{13}C enrichment therefore will be a necessary
condition for a sufficiently large S/N ratio in the spectra.

Fortunately, biosynthesis of plant viruses is a very
efficient and economic method to incorporate ^{13}C in the
viral protein, using $^{13}CO_2$ as a precursor. Tobacco Mosaic
Virus (TMV) and the TMV-protein have been prepared in this
way. Surprisingly, the linewidth broadening and lengthen-
ing of T_1 do occur in ^{13}C NMR spectra of TMV-protein at
80 KG, but to a lesser extent as predicted on the basis of
a rigid protein structure. In fact, in protein aggregates
T_1 and T_2 turn out to be almost independent of the molecular weight in the range from 10^5 to 10^7, suggesting an internal mobility in the TMV-protein. In the intact virus a
much lower internal mobility is observed. A discussion
will be given of the biological importance of the internal
mobility of coat proteins in virus assembly.

Reference

1. E. Oldfield, R.S. Norton and A. Allerhand (1975). *J. Biol. Chem.*
250, 6368–80.

10. THE STRUCTURE OF THE LYSINE-RICH HISTONES

G.E. CHAPMAN, F.J. AVILES, P.D. CARY, C. CRANE-ROBINSON,
P.G. HARTMAN and E.M. BRADBURY

*Portsmouth Polytechnic, Biophysics Laboratories,
St. Michael's Building, White Swan Road, Portsmouth*

The lysine rich (H1) histones are a class of basic proteins of about 21,000 MW, which occurs in large amounts
in eukaryotic chromosomes. Their detailed function is not
known, though it is generally agreed that they are involved in higher order structure in the chromosome. We have
shown, using ^1H NMR and hydrodynamic methods combined with
specific cleavage of the molecule that (in vertebrates at
least) it consists of three structural regions [1,2,3,4].
The central region of the sequence forms a globular structure in solution at high anion ionic strength, and this
globular structure is probably present in H1 in the chromo-

some. We have been studying the structure of this region and more recently the globular structure of a related protein, histone H5, from the chromosomes of nucleated erythrocytes. Our techniques involve principally ^{1}H NMR and theoretical secondary structure prediction. NMR techniques demonstrated will include (i) spin decoupling of perturbed peaks, (ii) effect of chemical modification on NMR spectra, (iii) complete assignment of histidines using differential deuteration and iodination, (iv) partial assignment of tyrosines by differential iodination and nitration, (v) assignment of resonances by melting (fast exchange denaturation) and (vi) titration studies.

References

1. Bradbury, E.M., Cary, P.D., Chapman, G.E., Crane-Robinson, C., Danby, S.E., Rattle, H.W.E., Boublik, M., Palau, J. and Aviles, F. (1975). *Eur. J. Biochem.* **52**, 605–13.
2. Bradbury, E.M., Chapman, G.E., Danby, S.E., Hartman, P.G. and Riches, P.L. (1975). *Eur. J. Biochem.* **57**, 521–8.
3. Chapman, G.E., Hartman, P.G. and Bradbury, E.M. (1976). *Eur. J. Biochem.* **61**, 69–75.
4. Hartman, P.G., Chapman, G.E., Moss, T. and Bradbury, E.M. (1977). *Eur. J. Biochem.* (in press).

11. STUDIES OF HISTONE COMPLEXES USING 270 MHz ^{1}H NMR

T. MOSS, H. HAYASHI, P.D. CARY, C. CRANE-ROBINSON and E.M. BRADBURY

Portsmouth Polytechnic, Biophysics Laboratories,
St. Michael's Building, White Swan Road, Portsmouth

The 4 histones H3, H4, H2A, H2B that make up the protein core of the nucleosome, form two well-defined cross-complexes in free solution — an $(H3/H4)_2$ tetramer and an $(H2A/H2B)_1$ dimer. The tetramer (which is capable of inducing considerable order in 140 b.p's of DNA in the absence of the other histones) has an NMR spectrum which although showing several perturbed aromatic and ring-current shifted peaks, does not have the overall chemical shift complexity typical for globular proteins. It is concluded that the tetramer is only partially globular and has mobile and unstructured N-terminal regions. This is substantiated by sedimentation velocity studies and by experiments on cross-complexation between large N and C-terminal peptides of both H3 and H4, using the characteristic perturbations in the aromatic spectrum as the criterion of complexation. It is found that complexation occurs when ~ 35 per cent of the N-terminal residues are absent from both H4 and H3, but not when the C-terminal 18 per cent of H4 is absent.

The H2A/H2B dimer likewise has a characteristic NMR spectrum and may also have free N-terminal chains. Conformational heterogeneity in the dimer is suggested by the

observation that trypsin treatment produces a limit-product
complex which lack N-terminal amino acids but which possesses all the NMR perturbations characteristic of the
native chain.

12. NMR INVESTIGATION OF HISTONE H1 PHOSPHORYLATION:
ITS EFFECTS ON THE ENTHALPY OF H1 FOLDING AND
ON THE FORMATION OF H1-DNA COMPLEXES

T.A. LANGAN, E.M. BRADBURY, S.E. DANBY and H.W.E. RATTLE

*Portsmouth Polytechnic, Biophysics Laboratories,
St. Michael's Building, White Swan Road, Portsmouth*

Histone H1 (molecular weight 21,000 daltons) is the largest of the five histone molecules found in eukaryotic
chromatin, and the only one not involved in the internal
structure of the nucleosome. It is thought to play a part
in the formation and modulation of higher-order folding of
the polynucleosomal fibre, and is subject to chemical modifications by specific enzymes, some of which are hormone
and cell-cycle dependent. Histone H1 phosphorylated at
serine-37, at serine-105, and at a number of sites associated with the prophase and metaphase stages of the cell
cycle has been investigated by NMR in isolation and complexed with DNA.

The region of H1 between residues 40 and 123 forms a
globular structure at pH > 5 and high ionic strength; one
effect of this structure formation on the NMR spectrum is
a chemical shift change in the aromatic resonances of
tyrosine-72 of about 0.2 ppm. Thermal denaturation of the
H1 is reversible, and a rapid exchange between structured
and unstructured forms takes place; in consequence the
chemical shift of the tyrosine resonances may be used to
calculate an equilibrium constant for the structure formation at different temperatures, and the enthalpy calculated via a van't Hoff plot. Phosphorylation at serine-105
causes a reduction from 25 ± 2 kcal mol^{-1} to 13 ± 2 kcal
mol^{-1}, and so provides a mechanism for control of the
structure of the globular region.

The carboxy-terminal half of H1 is thought to be the
site of interaction with internucleosomal DNA in chromatin:
NMR studies reveal major changes in the strength of complex formation at some ionic strengths on phosphorylation
of the H1 at only one or two sites, in particular the site
where phosphorylation is dependent on intracellular cyclic-
AMP levels and those associated with chromosome condensa-
tion during prophase. This provides an obvious mechanism
for at least part of the modulation of chromosome structure
through the cell cycle, and is the subject of further investigation.

13.   HIGH RESOLUTION NMR IN INTERSTITIAL SPACE:
STUDIES OF THE LYSOZYME-SOLVENT INTERACTION
IN A SINGLE PROTEIN CRYSTAL

MICHAEL K. CRAWFORD and PAUL C. LAUTERBUR

*Department of Chemistry, State University of New York
at Stony Brook, Stony Brook, New York 11794, U.S.A.*

Very general considerations suggest that mobile small
molecules within the interstitial regions of protein crystals should give high resolution NMR spectra showing partial orientation phenomena similar to those found for
solutes in liquid crystals. Usually, however, such spectra
will be greatly broadened by the field inhomogeneities
produced by macroscopic magnetic susceptibility effects,
even for diamagnetic proteins, and by the random orientations of the crystallites, which will give rise to broad
powder patterns. We have developed an apparatus that makes
possible the elimination of magnetic susceptibility broadening for single protein crystals while rotating them to
any desired orientation, and have observed oriented high
resolution proton NMR spectra for dilute methanol in a
single crystal of orthorhombic lysozyme, as well as orientation effects on the interstitial water proton resonance
and on the quadrupole splitting of the deuterium signal
of interstitial D_2O in the same crystals. This technique
offers a new opportunity for studying the interfaces between protein molecules and small molecules and solvents
under very precisely defined conditions.

14.   PULSE TECHNIQUES IN HIGH RESOLUTION
^{1}H NMR OF MACROMOLECULES

I.D. CAMPBELL, C.M. DOBSON and R.G. RATCLIFFE

*Departments of Biochemistry and Inorganic Chemistry,
South Parks Road, Oxford*

^{1}H NMR of macromolecules appears very attractive because
of the high sensitivity and ubiquity of the ^{1}H nucleus.
There are, however, three technical problems associated
with its use. The first is that the concentration of [1]H
in the solvent H_2O is very high compared to the macromolecule; the second is the small range over which the
resonances occur which leads to a resolution problem; the
third problem is associated with difficulties in assigning
the resonances to particular chemical groups. Methods
which alleviate these technical problems will be illustrated. These methods involve the use of pulsed double resonance techniques which reduce the solvent resonance and
identify spin coupled and exchanging resonances. In addition, methods will be described, which allow a spectrum of
a reduced number of resonances, a sub-spectrum, to be

obtained. It will also be shown how kinetic information
can be extracted using pulsed NMR.

15. ^{13}C NMR STUDY OF PROTEIN UNFOLDING

O.W. HOWARTH

Department of Molecular Science,
University of Warwick, Coventry, CV4 7AL

^{13}C NMR should be particularly well-suited for the study
of proteins in which some parts are effectively random-
coil and others structured, because the random-coil parts
will have their ^{13}C resonances selectively enhanced in
amplitude both by their reduced linewidth and also by a
greater nuclear Overhauser enhancement [1-3]. Thus it
should be an effective and detailed monitor of protein unfolding, requiring knowledge only of the primary structure
for interpretation.

The unfolding of RNase A is already fairly well understood [4], following a wide variety of investigations. It
therefore provides an ideal test for the above. The thermal
unfolding at two pH's has been investigated and the ^{13}C
spectra show that the unfolding sequence is indeed pH-
dependent, and fits well with the theory [4]. Considerable
progress has also been made in the simulation of the ^{13}C
spectra, both of structured and partially structured proteins, using independent shift and width parameters [2],
despite the uncertainties introduced by structure-derived
shifts. These simulations compare quite well with the observed spectra, especially in the β-carbon and aromatic
regions.

In addition, spectra of supposedly random-coil protein,
at 70°C, show residual structure associated with cysteine
and proline residues, which implicates these as nucleation
sites for refolding.

References

1. Clark, V.M., Lilley, D.M.J., Howarth, O.W., Richards, B.M. and
   Pardon, J.F. (1975). *Biochemistry* **14**, 4590.
2. Howarth, O.W. and Lilley, D.M.J. *Progress in NMR Spectroscopy* (in
   preparation).
3. Lilley, D.M.J., Howarth, O.W., Clark, V.M., Pardon, J.F. and
   Richards, B.M. (1976). *FEBS Lett.* **62**, 7.
4. Burgess, A.W. and Scheraga, H.A. (1975). *J. Theor. Biol.* **53**, 403.

16. ASYMMETRIC TITRATION CURVES OF THE HISTIDINE PROTONS
OF PHOSVITIN
INFLUENCE OF NEIGHBOURING PHOSPHOSERYL-RESIDUES

J. KREBS and R.J.P. WILLIAMS

*Lab Biochemistry III, ETH Zurich,
Universitätsstr 16, CH-8006, Zurich*

Phosvitin is a highly phosphorylated glyco-protein found
in egg yolk with a MW of about 35,000. More than 50 per
cent of the amino acids are phosphorylated serine-residues
which exist as clusters of up to 8 phosphoserine residues
in sequence [1]. We wished to discover by NMR measurements
if Phosvitin, reported to be a random coil protein [2],
can undergo local conformational rearrangements during a
pH-titration. From limited sequence data it is known that
most of the 11 Histidines of the protein are situated
nearby phosphoserine clusters [3]. The C2 and C4 protons
of histidine are sensitive to nearby titratable groups
especially with similar pK-values and therefore we followed the chemical shift changes of the proton signals of
the histidine residues during a pH titration. The following observations were made
(1) At pH < 5 the C2 and C4 protons respectively of all
histidines have the same chemical shift (random coil
spectrum)
(2) At pH ~ 7.5 one can differentiate between most if
not all C2 protons of the different histidines and
about the half of the different C4 protons by convolution difference spectroscopy.
(3) The titration curves are asymmetric.
(4) The apparent pK values are at least one pH unit
higher than those of the free histidine.
(5) Plots of log $(\delta_A - \delta)/(\delta - \delta_B)$ versus pH show strong
deviations from linearity and have different slopes.
A possible explanation for these observations is the
existence of hydrogen bond bridges between the histidines
and nearby phosphate groups which would give rise to local
conformational constraints. This view is supported by
Phosphorus NMR measurements which have been used as extensive checks of this conclusion. At high and low pH the
resonances are sharp but at intermediate pH very broad
lines are observed indicative of intermediate exchange
between constrained environments.

References

1. For review see Taborsky, G. (1974). *Adv. Prot. Chem.* **28**, 1.
2. Grizzuti, K.S. and Perlmann, G.E. (1970). *J. Biol. Chem.* **245**, 2573.
3. Belitz, H.D. (1965). *Z. Lebensm.-Unters. Forsch* **127**, 341; Shainkin,
R.S. and Perlmann, G.E. (1971). *J. Biol. Chem.* **246**, 2278.

17. PROTON MAGNETIC RESONANCE STUDIES OF PARVALBUMINS

A. CAVÉ, C.M. DOBSON, J. PARELLO and R.J.P. WILLIAMS

*C.N.R.S. E.R.140, U.S.T.L. 34060, Montpellier, France and
The Inorganic Chemistry Laboratory, South Parks Road,
Oxford OX1 3QR.*

Parvalbumins from muscle form a class of homologous proteins of low molecular weight (11,500). The proteins contain two calcium ions, have a high phenylalanine content
(8-10 residues) and a compact globular shape. Most of the
phenylalanine residues are internal and are conserved in
the sequence of different parvalbumins.

An analysis of the 270 MHz PMR spectra of parvalbumins
from several different species has been carried out. Most
of the aromatic proton resonances have been resolved, and
coupling patterns have been found using multiple resonance
techniques and spin echo pulse sequences. It has been
established that most, if not all, of the aromatic residues are undergoing rapid flips about the $C_\beta-C_\gamma$ bonds,
thus demonstrating the extensive internal mobility of
these proteins. Some evidence for more extensive conformational mobility has been collected.

The identification of the aromatic proton resonances of
phenylalanine residues has provided a method of comparing
the tertiary structures of the proteins from different
species. In addition, the effects on the tertiary structure of pH, dimerization and the removal of calcium have
been studied.

18. ASSIGNMENT AND STUDY OF TYROSINE AND TRYPTOPHAN RESIDUES OF LYSOZYMES

R. CASSELS, C.M. DOBSON, F.M. POULSEN and R.J.P. WILLIAMS

*The Inorganic Chemistry Laboratory, South Parks Road,
Oxford OX1 3QR.*

Resonances of the three tyrosines and the six tryptophan
residues of hen and bob white quail lysozymes have been
fully assigned in the PMR spectra.

The first stage of assignment involved resolution of
individual resonances, and assignment to amino acid type
The doublet resonances of all aromatic protons of each
tyrosine residue have been detected and spin decoupled.
The six C(2)H resonances of the tryptophan residues have
been identified from their singlet multiplet structure
which was detected by using spin echo pulse methods. The
only other singlet aromatic proton resonances (of the
single histidine residue) have also been assigned. The
N(1)H resonances of the tryptophans were identified in
the spectra of the lysozymes in H_2O solvent.

The second stage of assignment involved relating the

observed resonances to specific residues in the sequence.
This was achieved by perturbing the PMR spectra by specific chemical modifications; by binding of paramagnetic
probes (lanthanide ions and spin labels), inhibitors and
protons; and by solvent exchange studies. The experiments
showed that individual experiments, even chemical modifications, were usually insufficient to provide firm assignments. Instead, the results of a variety of experiments
had to be considered together.

In addition to allowing assignment of resonances, the
experiments have revealed interesting aspects of protein
structure in solution. These include detection of conformational mobility, and the effects on local protein conformation of chemical modification and ionisation of specific groups.

19. β-LACTAMASE II: INTERACTION OF HISTIDINE RESIDUES WITH THE ESSENTIAL Zn ATOM

G.S. BALDWIN,* H.A.O. HILL,[†] B.E. SMITH,[†] S.G. WALEY,*
and E.P. ABRAHAM*

*Sir William Dunn School of Pathology
[†]Inorganic Chemistry Laboratory
University of Oxford, Oxford OX1 3RE

β-Lactamase II, one of two β-lactamases produced by
Bacillus cereus, is a Zn-dependent enzyme (MW 22,000),
hydrolysing penicillins and cephalosporins. The enzyme has
been studied by 270 MHz proton NMR. pH titrations, Zn
titrations and histidine C2 proton exchange suggest that 2
of the enzyme's histidine residues are ligands for the Zn
atom involved in activity.

20. INTERNAL MOTIONS OF THE NEUROHYPOPHYSIAL HORMONES

L.J.F. NICHOLLS, C.R. JONES, J.J. FORD and W.A. GIBBONS

Department of Biochemistry, College of Agriculture and
Life Sciences, University of Wisconsin, Madison,
Wisconsin 53706, U.S.A.

A total PMR spectral analysis of approximately thirty residues in four different neurohypophysial hormones has been
performed using difference NMR and difference double resonance techniques. In several instances these studies have
been performed under conditions of varying temperature,
solvent, and pH. The data obtained from these studies provides considerable insight into the solution dynamics and
internal motions of these molecules.

If both cysteine Cα–Cβ rotamer populations determine
the corresponding Cβ–S populations and these themselves are
correlated through the S–S bond then there can be no more

than six possible conformers of the cyst*ine* moiety. This number is further reduced to three if the disulphide bridge has a single chirality. Our findings clearly support correlated motion for this whole unit.

Extensive internal rotations of the other residues of the neurohypophysial hormones is also consistent with our data. Further support for this motion is obtained by examining the temperature dependences of both coupling constants and chemical shifts. We have, for some of these molecules, examined the dependence on temperature of the chemical shifts of all (amide, aliphatic, and aromatic) proton resonances. In general the NH proton temperature dependencies are negative; the aliphatic protons, however, usually have temperature coefficients in the range ± 2 ppb/°C. These can be accounted for, in part, by postulating temperature independent intrinsic rotamer chemical shifts, and are an additional NMR parameter for investigating conformational dynamics. A further consequence of this interpretation is that greater care should be exercised in using these temperature coefficients to distinguish hydrogen bonded and non-hydrogen bonded amide protons especially when the value lies in the range -2 to -4 ppb/°C.

21. UNEQUIVOCAL DETERMINATION OF PEPTIDE STRUCTURE

A. FISCHMAN, D. COWBURN, D. LIVE, W.C. AGOSTA and
H. WYSSBROD*

The Rockefeller University, New York, N.Y. 10021
**Department of Physiology and Biophysics, Mount Sinai Medical
and Graduate Schools of the City University of New York*

We have sought methods for determining unequivocally the conformation of biologically active peptides, investigating both fixed conformations and those possessing rotational isomerism. Our principal technique is NMR measurements on specifically designed and synthesized isotopic isomers of natural amino acids and peptides, followed by conservative derivation of structural information from the available homo- and heteronuclear coupling constants.

Isotopic isomers are designed to (i) simplify coupled spin systems, (ii) reduce overlap of transitions in the spectrum, and (iii) enrich the less abundant isotopes of carbon and nitrogen (^{13}C and ^{15}N) that are needed for the measurement of heteronuclear coupling constants. NMR studies of isotopic isomers of the amino acid leucine demonstrate the feasibility of this approach, provide valuable calibration data and permit detailed calculation of the rotational isomerism of the leucine side chain. We are synthesizing isotopic isomers of the peptide hormone oxytocin (see below) with a view to applying this approach. Preliminary results include a description of the conformation of the disulphide bridge and of several side chains.

An analysis of the use of combinations of homo- and heteronuclear coupling constants in conformational analysis will be presented.

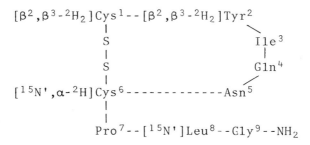

$$[\beta^2,\beta^3\text{-}^2H_2]Cys^1 \text{--} [\beta^2,\beta^3\text{-}^2H_2]Tyr^2$$

22. NMR OF NORMAL AND ^{13}C ENRICHED AMINOACIDS AND PEPTIDES

M. PTAK,* E. BENGSCH and A. HEITZ

Centre de Biophysique Moléculaire (CNRS) et Université d'Orléans
45045 Orléans Cedex, France*

A complete analysis (computer simulation) of the ^{13}C NMR spectra of uniformly enriched ($\simeq$ 85 per cent) Serine, Threonine, Aspartic acid and Glutamic acid is presented. The total spectra are reconstituted by addition of partial spectra corresponding to the different labelling probabilities. Such an analysis provides with a good accuracy all the $^{13}C\text{--}^{13}C$ coupling constants which are related to the structure and to the conformations of these aminoacids in aqueous solution. A comparison is made with $^{13}C\text{--}^1H$ and $^1H\text{--}^1H$ coupling constants measurements. Possible isotopic effects are discussed.

^{13}C enriched aminoacids are incorporated in short peptides in order to investigate the conformations of polar side chains and the interactions between them. As an example, the interactions between the Seryl (or Threonyl) and Histidyl residues are studied by NMR.

NMR and crystallographic studies as well as conformational calculations performed on a number of linear and cyclic dipeptides reveal the existence of specific interactions between polar side chains and a strong influence of the hydration.

On these bases, an explanation of the catalytic properties of short peptides containing Ser, His and Asp residues can now be considered.

BILE SALT AGGREGATION

23. J. ULMIUS,* T. WIELOCH,*† S. FORSÉN,* and B. BÖRGSTROM†

*Department of Physical Chemistry II and
†Department of Physiological Chemistry,
University of Lund, Lund, Sweden

Bile salts are physiological active detergents playing an important role in the degradation process of lipids in the intestinal lumen. They differ in many aspects from other detergents, e.g., their aggregation properties. We have specifically labelled sodium deoxycholate, sodium cholate, and sodium chenodeoxycholate with deuterium (Fig.) and studied the linewidth changes of the deuterium signal in a concentration range of 1-150 mM with deuterium NMR.

Four distinct plateaus could be observed at which the linewidth increase levels off. The concentration at which these plateaus appear and the amplitude of the linewidth change is dependent on the number and the position of hydroxyls in the molecule and the position of the label.

$(T_2)^{-1}$ for deuterium is solely dependent on the correlation time τ_c and the linewidth is thus a measure of the size of the aggregate in which the deuteron is positioned.

We interpret the observed plateaus as being due to formation of stable aggregates of a definite size. The size and aggregation pattern of the bile salts will be discussed.

24. BINDING OF CARBON MONOXIDE TO HEMOPROTEINS AND MODEL HEME-COMPLEXES AS STUDIED BY ^{13}C NMR

R. BANERJEE and J.M. LHOSTE

Institut de Biologie physico-chimique, Paris
Institut du Radium, Orsay

Chemical shifts for the ^{13}C of the carbonyl ligand bound to hemoproteins are used as probes for metal-ligand interactions and of possible second order effects of the heme environment. Data obtained on a number of carbonyl-hemoproteins are compared with known reaction parameters including the stability constants of the carbonyl complexes. A one ppm upfield shift of the ^{13}C resonance corresponds to an increase of about 1 kcal M^{-1} per site for the CO-binding reaction.

In order to estimate the contribution of the *trans* effect of the axial base to the different chemical shifts, experiments have been performed on model ternary compounds,

each having a nitrogenous base and a carbon monoxide molecule liganded at the two axial positions of the ferrous
iron of the protoheme. In most of the ternary complexes,
the carbonyl ligand exchanged slowly with the bulk carbon
monoxide whereas the nitrogenous base, used in very large
molar excess, underwent rapid exchange. The nitrogenous
bases examined belong to several types: pyridinic (pyridine
and substituted pyridines), aromatic and aliphatic primary
and secondary amines (aniline and substituted anilines,
quinolines, piperidine, n-alkyl to *tert*-alkyl amines) and
imidazoles. The observed chemical shifts are considered in
relation to the basicity of the nitrogen ligand and substituent effects which include, for some cases, steric
factors.

In other experiments where carbon monoxide was used
together with a mixture of two suitable chosen bases, a
single broader resonance was observed. Consideration of
chemical shift values for such systems lead to the conclusion that an increasing affinity of the base for the reduced heme can lead to a high field shift of ^{13}C resonance of the carbon monoxide within a range of 3 to 4 ppm.

25. PROTEIN INTERACTIONS WITH GENERAL ANAESTHETICS

F.F. BROWN,* M.J. HALSEY[†] and R.E. RICHARDS*

*Department of Biochemistry, South Parks Road, Oxford
†Division of Anaesthesia, Clinical Research Centre,
Northwick Park, Harrow, Middlesex

This programme comprises three parts: the interactions
with (1) a model protein (haemoglobin), (2) a functionally affected protein (luciferase), (3) a neuro-protein
(acetylcholine receptor). We are exploring the hypothesis
that anaesthetic actions can be explained on the basis of
direct interactions with proteins in addition to lipid and
aqueous effects. However, no one has demonstrated that
hydrophobic pockets in proteins behave like a bulk phase
in terms of anaesthetic solubility, nor is there any firm
evidence for direct protein perturbations which can be
correlated with anaesthetic potencies.

Our results from model protein studies have shown that
general anaesthetics can produce specific perturbations at
clinical concentrations (Brown *et al.* (1976). *Proc. Roy.
Soc.* B **193**, 387) and that different specific perturbations
are produced by different anaesthetic agents (Barker *et
al.* (1975). *Brit. J. Anaesth.* **47**, 25). More recently we
have observed that the relaxation enhancement produced in
the most shifted resonances appears to be a maximum at
approximately the same molecular concentration independent
of the type of anaesthetic. A maximum linewidth of a perturbed resonance should be observed when approximately 40
per cent of the binding sites causing the perturbation are

filled (cf. Baldo *et al.* (1975). *Biochem.* **14**, 1893). A
binding site in this context may be thought of as a preferred hydrophobic environment within the molecule. Several types of general anaesthetic have been studied using
Fourier Transform spin-echo techniques to amplify the small
changes in T_2 observed.

It is interesting to note that the observed specific
perturbations actually occur in non-hydrophobic parts of
the molecule (e.g. titratable histidine C2 resonances at
the protein surface). This correlation with lipid solubility and with anaesthetic potency is an encouraging asset
for any mechanism of anaesthesia involving protein perturbations.

26.   THE BINDING OF NITROAROMATIC HAPTENS
TO THE MOUSE MYELOMA PROTEIN 315

S.K. DOWER, R.A. DWEK, P. GETTINS, R. JACKSON,
S. WAIN-HOBSON and D. GIVOL*

*Department of Biochemistry, South Parks Road, Oxford
*Department of Chemical Immunology
The Weizmann Institute of Science, Rehovoth, Israel*

The binding site for the 2,4-Dnp binding mouse myeloma IgA
protein 315 has been studied by a combination of model
building and magnetic resonance methods. This has led to
a picture of the combining site contact residues at high
resolution (Dwek, R.A., Wain-Hobson, S., Dower, S.,
Gettins, P., Sutton, B., Perkins, S.J. and Givol, D.
(1977). *Nature* **266**, 31). The work reported here extends
the study to the binding of 2,4,6 Tnp derivatives and
2,4,-Dinitronaphth-1-ol-7-sulphonic acid. Magnetic resonance studies, in particular high resolution proton NMR,
have shown that these haptens stack with Trp 93_L in the
aromatic box formed by Phe 34_H and Tyr 34_L, and that the
mode of binding of the hapten aromatic ring within the
aromatic box is very similar to that of Dnp derivatives.
It has also been shown by a transfer of saturation experiment with ε-N-Tnp-α-N-acetyl-Tnp-L-lysine that the mode
of binding of this hapten is very similar to that of Tnp-
L-aspartate despite a difference in three orders of magnitude in binding constants.

27. AN NMR STUDY OF DRUGS DESIGNED TO FIT A RECEPTOR SITE OF KNOWN STRUCTURE

F.F. BROWN and P.J. GOODFORD

*Department of Biochemistry, South Parks Road, Oxford
Department of Biophysics and Biochemistry,
Wellcome Research Laboratories, Beckenham, Kent*

Bis-arylhydroxysulphonic acids have been designed to inter-act with the 2,3-diphosphoglycerate (DPG) binding site in haemoglobin (Beddell *et al.* (1976). *Br. J. Pharmacol.* **57**, 201). These compounds matched 2,3-DPG in size and were expected to interact covalently with the terminal amino groups in the β-β cleft. Oxygen binding studies on haemo-globin in the presence of these compounds revealed the anticipated displacement of the O_2-dissociation curve but did not give any direct evidence that the mechanism and binding site were, in fact, the same as for 2,3-DPG.

NMR studies on haemoglobin have revealed a downfield shift in the β2 and β143 histidine C2 resonances, charac-teristic of the interaction of a negatively charged species in fast exchange at the binding site. The sulphonic acids caused a collapse of the same resonances. This effect was most marked with the compounds which had been found to give the greatest shift of the O_2-dissociation curve. The result was further confirmed by competition studies between DPG and the most effective of the sulphonic acids. Here, the DPG was used to shift the resonance lines away from the other peaks in the spectrum and the sulphonic acid was then observed to collapse only these resonances.

These results provide convincing evidence that the drugs compete with 2,3-DPG for the same site. They also show that the drugs are in slow exchange, which supports the sugges-tion that the compounds bind covalently.

28. X-RAY AND NMR STUDIES ON HEN EGG WHITE LYSOZYME

S.J. PERKINS, L.N. JOHNSON, D.C. PHILLIPS and R.A. DWEK*

*Laboratory of Molecular Biophysics, Zoology Department,
South Parks Road, Oxford OX1 3QU
Department of Biochemistry, South Parks Road, Oxford

To develop the conclusions that are drawn from the use of one technique, independent determinations of detailed bio-logical structures are essential. NMR and X-ray studies will be reported for the enzyme-inhibitor complexes of lysozyme with N-acetyl-D-glucosamine and its α and β gly-cosides.

The crystallographic data after real space refinement for tetragonal crystals of lysozyme-inhibitor complexes have been used to make ring current predictions for the six subsites of lysozyme. From the fully bound chemical shifts of the sugar inhibitors binding to lysozyme, the monomer binding sites in solution are directly identified as subsites C and E from the NMR evidence. The crystal structures show that the conformational changes in the protein are rationalized by postulating a flexible structure to the left of the active site cleft and a relatively more rigid structure to the right.

Crystallographic studies on tetragonal lysozyme-Gd(III) complexes show two Gd(III) positions in one major site with coordinates similar to those for triclinic lysozyme-Gd(III) complexes. The binding is monodentate to each of Asp 52 and Gly 35 and results from the low solvent accessibility, the rigid positioning of Asp 52 and Gly 35 in the protein and the presence of Asn 57 between these carboxyls.

NMR studies on lysozyme-lanthanide inhibitor complexes show that the inhibitor competes with Ln(III). This arises from competition at subsite E (and the anomers at subsites C-α also). The limitations of lanthanides in paramagnetic mapping of these complexes are discussed.

29. EFFECT OF METHANOL AND ANILINE ON Zn^{++} AND Co^{++}-CARBONIC ANHYDRASE. ^1H AND ^{13}C NMR STUDIES

R. DESLAURIERS,* I.C.P. SMITH,* J. WESTERIK[†] and B. SARKAR[†]

*Division of Biological Sciences, National Research Council of Canada, Ottawa, Canada K1A OR6
[†]Research Institute, The Hospital for Sick Children, Toronto, Ontario, Canada M5G 1X8

Methanol and aniline inhibitors have been used to probe the active centre of carbonic anhydrase. Aniline and methanol specifically and reversibly inhibit the enzyme with a Ki of 0.022 M and 0.9 M, respectively. Methanol inhibition increases with increasing pH whereas aniline inhibition decreases at pH values greater than 7. The methanol inhibition follows the enzymic pH-activity profile. Both visible and NMR spectroscopic studies have been performed on the Co^{++}-substituted enzyme. Aniline and methanol are non-competitive with respect to substrate but bind to the enzyme competitively. Aniline, but not methanol, was found to alter the characteristic absorption spectrum of Co^{++}-carbonic anhydrase.

^1H NMR of H_2O in solutions containing Co^{++}-carbonic anhydrase was performed in order to test whether the inhibition by aniline and methanol could be associated with replacement or perturbation of the water molecule which is thought to occupy the fourth coordination position on the metal in the enzyme. At pH 8.2 and 7.2, neither methanol nor aniline perturbs the T_1 values observed for H_2O in the presence of Zn^{++}- or Co^{++}-carbonic anhydrase. KCN did however perturb the T_1 values of H_2O in similar solutions. At lower pH the T_1 values of H_2O are longer than at 8.2, in accord with data of Fabry *et al.* (*J. Biol. Chem.* **17**, 4256 (1970)) on Co^{++}-carbonic anhydrase and Lanir *et al.* (*Biochemistry* **14**, 242 (1975)) on Mn^{++}-carbonic anhydrase.

^{13}C T_1 measurements were performed using 90 per cent ^{13}C-enriched methanol. No line width perturbations of the ^{13}C resonance of methanol was observed in the presence of apoenzyme, Zn^{++}- or Co^{++}-carbonic anhydrase. T_1 values were sensitive to the presence of Co^{++}-carbonic anhydrase. Acetazolamide and NaCN did not perturb ($<$ 6 per cent) the T_1 values observed for methanol in the presence of the Co^{++}-enzyme at pH 7.2. Aniline and KCN were both without effect on the T_1 values at pH 8.2. A distance of *ca* 6 Å was calculated between the paramagnetic centre and the ^{13}C atom of methanol.

The above studies show that methanol does not bind directly to the Co^{++} ion of carbonic anhydrase and does not perturb the water bound to the metal ion. At pH 7.2 CN^- perturbs the water relaxation times whereas aniline does not. The visible absorption spectrum of the Co^{++}-enzyme is altered by aniline suggesting that aniline perturbs the ligand environment and perhaps causes slight local conformational changes. Competitive inhibition occurs between aniline and methanol indicating overlapping (or mutually exclusive) binding areas of the active site.

30. ANION BINDING SITES OF PROTEINS QUADRUPOLE RELAXATION STUDIES OF SYMMETRIC IONS

P. REIMARSSON, T.E. BULL, T. DRAKENBERG, B. LINDMAN and J.-E. NORNE

Physical Chemistry 2, Chemical Centre P.O.B. 740, S-220 07, Lund, Sweden

The interaction of anions with proteins has been subject to intensive studies by quadrupole relaxation in recent years. The $^{35}Cl^-$ relaxation has been found to give valuable information concerning the stoichiometry and binding affinity of small ions and other ligands to proteins. Also the correlation times and quadrupole coupling constants for Cl^- in different binding sites on the protein can be obtained and these give information concerning the mobility and type of interaction at the binding site. Discrimination between metal-coordinative and general anion binding sites can be achieved by the use of complex anions in competition studies using $^{35}Cl^-$ relaxation. Another way to study non-metal binding sites in proteins is by the quadrupole relaxation of a nucleus residing at a site of cubic symmetry in ions that do not coordinate to metal ions. Examples are $^{13}ClO_4^-$ and $^{75}AsF_6^-$. Here the quadrupole coupling constant may be affected by the distortion of the ion, as for the unperturbed ion the electric field gradients cancel. The ^{35}Cl quadrupole relaxation of the perchlorate ion bound to human serum albumin has given promising results. We have also studied in some detail the ^{19}F NMR spectrum of AsF_6^- bound to human serum albumin. The ^{19}F spectrum consists of four lines arising from spin-spin coupling with the ^{75}As nucleus which has a spin quantum number $I = 3/2$. For the pertaining non-extreme narrowing situation it is possible to fit the experimental spectra to theoretical ones by using two relaxation rate constants. From these constants the correlation time and the quadrupole coupling constant for the bound AsF_6^- ion can be calculated. It has thus been found that the correlation times for Cl^-, ClO_4^- and AsF_6^- are 11.2 ns, 20 ns and 30 ns respectively. As they according to competition experiments bind to the same sites, rapid internal mobility at the anion binding sites of serum albumin is indicated.

Surveying studies of the perchlorate ion in the presence of some other proteins will also be presented and the perchlorate ion is found to be a potential general probe of anion binding sites in proteins providing information on mode of binding and microdynamics.

31. TRIOSE PHOSPHATE ISOMERASE: INTERACTIONS WITH LIGANDS STUDIED BY ^{31}P NUCLEAR MAGNETIC RESONANCE

I.D. CAMPBELL,* P.A. KIENER,[†] S.G. WALEY[†] and R. WOLFENDEN[‡]

*Department of Biochemistry and
[†]Sir William Dunn School of Pathology, University of Oxford
[‡]Department of Biochemistry, University of North Carolina,
Chapel Hill, North Carolina 27514, U.S.A.

The ^{31}P nuclear magnetic resonance spectra of the substrate, dihydroxyacetone phosphate, and the two inhibitors, glycerol-3-phosphate and 2-phosphoglycollate, have been observed in the presence of triose phosphate isomerase. Whereas phosphoglycollate only bound as the trianion, dihydroxyacetone phosphate and glycerol-3-phosphate bound either as the singly or doubly charged species.

32. WATER PROTON RELAXATION AS A PROBE OF MOLECULAR
MOTION IN PARAMAGNETIC ION. MACROMOLECULE COMPLEXES

D.R. BURTON, S. FORSEN, G. KARLSTROM,
R.A. DWEK, A.C. McLAUGHLIN and S. WAIN-HOBSON

*Department of Pyysical Chemistry 2, Chemical Centre,
Lund Institute of Technology, Lund, Sweden
Department of Biochemistry, University of Oxford, Oxford U.K.*

Proton Relaxation Enhancement (PRE) — the observation of
an enhanced relaxation of solvent water protons in the
hydration sphere of a paramagnetic ion as a result of the
ion binding to a macromolecule — is an attractive starting
point for the study of macromolecules because of the high
concentration of water in aqueous solutions. However,
whilst the use of the method as a titration indicator is
straightforward, the same cannot be said of its use in the
field of micromolecular dynamics where the parameters of
molecular motion are extracted from a multiparameter best-
fitting procedure of theory to experimentally obtained relaxation data at varying frequency and temperature. As
many of the parameters involved are interdependent large
uncertainties might be expected in their values from an
ability to vary in a mutually compensatory manner to yield
a number of theoretical fits to the experimental data which
are equivalent or nearly so. Such complications have been
neglected when workers in the field have used PRE analyses
to make firm biological conclusions. For example the method
has even been used to propose schemes of enzyme mechanism.
   Using PRE we have studied complexes of Gd(III) with
immunoglobulin G(IgG) and a number of antibody fragments.
Despite the accumulation of relaxation data over a wide
range of frequency and temperature, computer analysis
showed that many of the molecular motion parameters govern-
ing the solvent water relaxation behaviour were poorly defined. In particular we found that the coordination number
of the metal ion (q) was ill-defined when there was a size-
able contribution to the relaxation rates from the exchange
of water molecules between metal ion and bulk solvent
('slow exchange contribution') e.g. q for IgG. Gd(III)
could take values between 2 and 8. In contrast the coordination number was found to be relatively well-defined
for systems in which there was no slow exchange contribution and it was found possible to determine q from a single
measurement of T_1 and T_2 at a frequency such that the product of the Larmor frequency (ω_I) and the correlation time
for the dipolar relaxation processes (τ_c) was approximately
unity.
   For the IgG-Gd(III) complex a rotational correlation
time (τ_R) was obtained which suggested considerable internal motion of the Fc region (C-terminal half of heavy chain
dimer) of the IgG molecule but large errors made unequivocal conclusions impossible. In view of the significance of

the Fc region in complement activation, it was clearly
important firmly to establish the existence of the motion
described. With this in mind we designed a novel approach
to PRE based on solvent proton-deuteron comparative measurements at a limited number of magnetic fields.

As compared to conventional PRE the method is much sim-
pler to apply (exact solution of equations rather than
multiparameter best-fitting is involved), the errors in
the derived parameters are smaller and the amount of data
required for unambiguous interpretation is much less. We
were able to use the method to show that there is indeed
considerable internal motion in the Fc region and investigations of the significance of the motion with respect to
the trigger of complement are underway.

33. BINDING OF HYDROGEN IONS, METAL IONS AND ANIONS TO TRANSFERRIN AND CONALBUMIN AS REVEALED BY HIGH RESOLUTION PROTON MAGNETIC RESONANCE

R.C. WOODWORTH, R.J.P. WILLIAMS and R.M. ALSAADI

*Inorganic Chemistry Laboratory, Oxford University,
South Parks Road, Oxford OX1 3QR*

The iron-binding siderophilins, transferrin from human
plasma and conalbumin from hen's egg white, have been
studied by proton magnetic resonance at 270 MHz with a
view to discovering the details of specific metal ion and
anion binding to the 2 sets of binding sites in these proteins. The spectra show distinct shifts in the resonance
positions of the C(2)-H's of certain histidyl residues on
titration of the protein with hydrogen ions, with anions,
or with metal ions in the presence of anions. The reported
requirement for synergistic binding of anions in order for
metal ion binding to occur is corroborated by our experiments. The enhancement of the relaxation rate of bulk
water by praseodymium ion is not affected by conalbumin
alone, but is suppressed by conalbumin plus the anion
2,5-dipicolinate (dipic) until two equivalents of praseodymium have been added per mole of protein. The binding
of 2 moles of 1:1::Pr:dip: per mole of protein reveals a
spectrum for bound dipic similar to that for the 1:3::Pr:
dipic complex. Excess 1:1::Pr:dipic appears in the expected region of the spectrum. One can conclude that the
metal ion is bound close to the anion on the protein and
that the ternary complex is in slow exchange.

34. SODIUM-23 NMR AS A PROBE FOR SELF-ASSOCIATION
OF A GUANOSINE NUCLEOTIDE

PIERRE LASZLO and AGNÉS PARIS

*Institut de Chimie, Université de Liège,
Sart-Tilman par 4000 Liège 1, Belgium*

5'guanosine monophosphate self associates into highly
ordered structures in aqueous solution: gels form at
acidic pH [1]. They may have played a role in the prebio-
tic appearance of polynucleotides [2]. In neutral or
slightly alkaline solutions, it forms planar tetramers
which further stack into an helical array [3].

Linewidth of sodium-23 NMR can serve as a probe for
molecular aggregation. We observed pronounced line broadenings of the ^{23}Na linewidth as a function of 5--GMP concentration between 0 and 0.8 H and pH 8.0. Other nucleotides (5'-AMP, 2'-GMP), in the same conditions, do not
display ^{23}Na NMR line broadenings of comparable magnitude.

From these results, we deduce new important information: the thermodynamic parameter for the aggregation process, the correlation time for the aggregate. Melting
curves are also obtained and characterized by T_m values
and cooperativity which are in excellent accord with those
parameters obtained earlier from IR[4].

References

1. Gellert, M., Lipsett, M.N., Davies, D.R. (1962). *Proc. Nat. Acad. Sci. U.S.A.* **48**, 2013.
2. Sulton, J.E., Lohrmann, R., Orgal, L.E., Miles, H.T. (1968). *Proc. Nat. Acad. Sci. U.S.A.* **60**, 409.
3. Pinnavaia, T.J., Miles, H.T., Becker, E.D. (1975). *J. Am. Chem. Soc.* **97**, 7198.
4. Miles, H.T., Frazier, J. (1972). *Biochem. Biophys. Res. Comm.* **49**, 199.

35. NMR STUDY OF THE HYDRATION WATER OF DNA AND
POLYNUCLEOTIDES USING FROZEN SOLUTIONS

R. MATHUR-DE VRÉ

*Universite Libre de Bruxelles,
Laboratoire de Biophysique et Radiobiologie*

Water associated with DNA and polynucleotides (hydration
water) is the non-freezable fraction at temperatures below
the freezing point of solvent, and is responsible for the
observed water proton and deuteron signals from the frozen
solutions. The sharp and well-defined water proton signals
from frozen H_2O and H_2O/D_2O solutions show a marked sensitivity to changes in the structure of macromolecules and
the effects of γ-irradiation on DNA and poly (A + U) complex. Striking modifications in the hydration water spectra

that result after irradiating the solutions at 0°C and
-80°C, or molecular association of polynucleotides can be
attributed mainly to the effects of proton transfer in the
hydrogen bonded path of the hydration layer formed by H_2O
molecules and the hydrogen bonding groups of the macro-
molecular chains. A comparison of the results obtained
from H_2O and D_2O solutions, respectively, as well as the
temperature-dependent behaviour of the linewidth of proton
signal in X%H_2O/Y%D_2O solutions of DNA illustrate clearly
the importance of proton transfer effects in frozen solutions.

Furthermore, several examples show that detailed information related to the influence of varying the structure
of macromolecules on their hydration characteristics can
be studied by NMR much more effectively in the frozen
state than at temperatures above 0°C.

36. RELATION BETWEEN BASE-STACKING AND BACKBONE
CONFORMATIONAL PROPERTIES OF DINUCLEOSIDE PHOSPHATES

D.B. DAVIES

*Department of Chemistry, Birkbeck College,
Malet Street, London WC1E 7HX*

The sugar phosphate backbone conformational properties of
mononucleotides, dinucleoside phosphates, a dinucleotide
and a polynucleotide have been analysed from recent highfield ^1H NMR measurements. The O(5')-C(5') and C(5')-C(4')
bond conformations of monomers depends on the base-ring
syn-anti equilibrium and changes in furanose ring conformations; this behaviour is reflected by results for dimers.
It is found that the sugar phosphate backbone conformations* of both dipurine and dipyrimidine dinucleoside phosphates exhibit linear correlations with the proportion of
based-stacked conformations. At zero base-stacked conformations differences in behaviour between purine and pyrimidine derivatives are found for dimers; this behaviour
is similar to that observed for monomers and reflects
differences in *syn-anti* equilibria of the base rings.
[The latter phenomenon is being investigated by analysis
of $^3J(^{13}C,H_1)$ magnitudes observed in proton coupled ^{13}C
spectra.]

The results also show that the highly stable backbone
conformational unit (akin to the 'rigid' nucleotidyl unit)
is given by *ca* 50-55 per cent contributions of base-stacked
conformations for dipurine dinucleoside phosphates whereas
the larger contribution needed for dipyrimidines (*ca* 65-
75 per cent) reflects the expected effect that base-
stacking interactions between purine base-rings are
greater than between pyrimidine rings. The few available
results for dideoxyribonucleoside phosphates exhibit similar behaviour to their diribonucleoside phosphate ana-

logues. The few results for heterodimers exhibit conformational properties that depend on the nature of the -pY
fragment. It is suggested that the relation between base-
stacking and backbone conformational properties of dinucleoside phosphates provides a method for investigating
the strength of base-stacking interactions of homodimers
and indicates how results for heterodimers might be
rationalised.

---

*Expressed in terms of both $\Sigma(=J_{4'5''} + J_{4'5''})$ and $\Sigma(=J_{P5'} + J_{P5''})$.

37. A NUCLEAR MAGNETIC RESONANCE STUDY OF QUINACRIN-NUCLEIC ACID INTERACTION

A.L. ANDREWS, A. SIMMONDS and K. WILSON

*Chemical Sciences, School of Natural Sciences,
The Hatfield Polytechnic, P.O. Box 109, College Lane,
Hatfield, Herts. AL10 9AB*

There is disagreement whether the binding of the anti-
malarial quinacrin to DNA depends on base composition of
the nucleic acid or is non-specific. Fluorescence and
exciton splitting of U.V. spectra suggest preferential
binding to A, T-rich regions, but association constants
derived from spectroscopy or ultra centrifugation indicate
non specific interaction with all bases.

We have used NMR to elucidate the nature of this binding, using mono- and dinucleotides as model nucleic acid
systems.

The NMR spectrum of quinacrin has been studied as a
function of pH and concentration, and an assignment made
with the help of spin-decoupling experiments. The six
protons of the quinacrin base have been assigned as two
overlapping ABC systems, and an excellent least squares
fit obtained.

This assignment has then been used to study the binding
of purine nucleotides to quinacrin base. In all cases preferential interaction with the chlorine containing ring
is found. Evidence for hydrogen bonding to the acridinium
proton is also forthcoming, but plots of induced shift vs.
concentration fail to reveal significant differences between the purine nucleotides. It is hence concluded that
quinacrin binding is non-base specific. A possible structure of the complex is presented.

38. TRANSBILAYER DISTRIBUTIONS AND MOVEMENTS OF
LYSOPHOSPHATIDYLCHOLINE (LPC) AND PHOSPHATIDYLCHOLINE (PC)
IN VESICLES AS MEASURED BY ^{13}C-NMR USING [N-^{13}CH$_3$]
LABELLED LIPIDS

B. de KRUIJFF

*Institute of Molecular Biology, University of Utrecht,
Padualaan 8, Utrecht, The Netherlands*

The outside-inside distribution of LPC and PC in mixed
vesicles was measured using Dy^{3+} as a paramagnetic ion
which cannot penetrate the vesicles and only can interact
with the outside facing lipid molecules causing a downfield
shift of the [N-^{13}CH$_3$] resonance.

LPC is preferentially localized in the outside monolayer of the vesicle. The transmembrane movement of LPC
was found to be an extremely slow process with a half time
excess of several days.

Using a pure phosphatidylcholine exchange protein
[N-^{13}CH$_3$] labelled PC could be introduced in the outer
monolayer of non labelled PC vesicles. The inward flow of
labelled PC was found to be extremely slow in vesicles
composed of one PC species. Under certain conditions a
relative fast transbilayer movement of PC in mixed vesicles was observed.

39. ^{1}H-NMR STUDIES ON THE INTERACTION OF CANNABINOIDS
WITH SONICATED EGG YOLK LECITHINS AND SARCOPLASMIC
RETICULUM MEMBRANES

R.J.J.Ch. LOUSBERG

Organic Chemistry Laboratory, University of Utrecht, Holland

Since recently, cannabinoids and analogues are frequently
studied for their possible useful application as e.g. a
psychiatric aid. The characteristic hemp constituent for
psychotropic action, Δ1(2)-tetrahydrocannabinol (THC),
differs in structure from other centrally active compounds
in not having a basic N-atom and being a neutral, a-polar
substance.

A possible suggestion for the mode of action of THC at
the molecular level is through its interaction with phospholipid components of the nerve cell membrane. At the
same time the question can be raised whether a THC-specific
receptor does exist.

The interaction of THC and cannabidiol (CBD) with phospholipid systems was first studied by ^{1}H-NMR using sonicated egg yolk lecithins (EYL), vesicles. Thus it was concluded by the very strong and selective line-broadening of
the methylene envelope that both THC and CBD show strong
hydrophobic interaction at the model membrane level. Subsequently, the interaction of THC with an isolated bio-

logical membrane was studied, using sarcoplasmic reticulum
(SR).

SR is a biological system still possessing functional
properties like Ca^{2+}-uptake and ATP-hydrolysis. Due to the
relatively low cholesterol content and high percentage of
unsaturated fatty acids the system is particularly suitable for ^{1}H-NMR studies. The appearance of a substantial
N-Me choline signal in the spectrum can be achieved by e.g.
temperature-raise, trypsin-treatment and sonication resulting in a change in the integrity of the protein structure
and hence elimination of the original lipid-protein interaction causing subsequent strong intermolecular dipolar
broadening of the N-Me choline signal.

Interaction of THC with SR does cause disruption of the
integrity of the lipid-protein structure.

40. MAGNETIC RESONANCE STUDIES OF A PROTEIN
IN A BACTERIAL MEMBRANE

N. LEE,*[†] P.C. LAUTERBUR* and M. INOUYE[†]

*Department of Chemistry and [†]Department of Biochemistry,
State University of New York at Stony Brook,
Stony Brook, New York 11794, U.S.A.

The envelope of a Gram-negative bacterium such as *E. coli*
is composed of an outer membrane, an inner or cytoplasmic
membrane, and a peptidoglycan network located between the
two membranes. The most abundant protein in *E. coli* is a
lipoprotein [1-3] of about 7200 daltons that is found only
in the outer membrane, and is present there in two forms.
In each cell, about 2.5×10^5 molecules of this lipoprotein are covalently bound, through the e-amino group of
the carboxyl-terminal lysine residue, to the peptidoglycan
layer, and about 5×10^5 otherwise identical molecules are
not covalently bound [4—9]. The complete sequence of the
protein, and the structure of the unusual amino-terminal
lipid group, are known, but its form and function in the
membrane are uncertain. It has been suggested that the
free and bound forms are assembled together to form super-
helical tubular hydrophilic channels, anchored at one end
to the peptidoglycan and extending through the outer membrane, and that these channels permit the passive transport of water-soluble molecules through the outer membrane
[10]. We are studying the NMR and ESR spectra of ^{13}C, ^{19}F
and spin-labelled forms of the lipoprotein in membrane
fragments and in intact bacteria. At this time, we have
evidence for a moderate amount of mobility of a tyrosine
side-chain in the carboxyl-terminal region, but no evidence
that the bound and free forms give rise to different spectra. This observation is not inconsistent with the idea
that both forms are components of the same rigid structure
[10]. NMR studies of solutions of the free form of the

lipoprotein [11] are also in progress.

References

1. Braun, V. and Bosch, V. (1972). *Proc. Natl. Acad. Sci. U.S.A.* **69**, 970.
2. Braun, V. and Bosch, V. (1972). *Eur. J. Biochem.* **28**, 51.
3. Hantke, K. and Braun, V. (1973). *Eur. J. Biochem.* **34**, 284.
4. Inouye, M., Shaw, J. and Shen, C. (1972). *J. Biol. Chem.* **247**, 8154.
5. Hirashima, A., Wu, H.C. Venkateswaran, P.S. and Inouye, M. (1973). *J. Biol. Chem.* **248**, 5654.
6. Braun, V. and Sieglin, J. (1970). *Eur. J. Biochem.* **13**, 336.
7. Braun, V. and Wolff, H. (1970). *Eur. J. Biochem.* **14**, 387.
8. Bosch, V. and Braun, V. (1973). *FEBS Lett.* **34**, 307.
9. Lee, N. and Inouye, M. (1974). *FEBS Lett.* **39**, 167.
10. Inouye, M. (1974). *Proc. Natl. Acad. Sci. U.S.A.* **71**, 2396.
11. Inouye, S., Takeishi, K., Lee, N., DeMartini, M., Hirashima, A., Inouye, M. (1976). *J. Bacteriol.* **127**, 555.

41. ^{31}P NMR OF PERFUSED WORKING RAT HEARTS

D.P. HOLLIS, W.E. JACOBUS, R.L. NUNNALLY,
G.J. TAYLOR, IV, and M.L. WEISFELDT

Johns Hopkins School of Medicine, Baltimore, Md. 21131

Recently, ^{31}P NMR has been used to analyse excised muscles for several phosphate-containing compounds, including the sugar phosphates, glycolytic intermediates, free ortho-phosphates, phosphocreatine, ADP and ATP. The intracellular pH of intact muscles was also determined. These muscles were neither perfused nor functioning. We will report the assessment by ^{31}P NMR of the energy status of the working, perfused rat heart. The ^{31}P spectrum of active muscle is characterised by prominent resonances from the three phosphates of ATP and the phosphate group of phosphocreatine, and only weak resonances from inorganic phosphate and sugar phosphates. Partial ischemia caused the intensities of the ATP and phosphocreatine resonances to decrease relative to that of P_1. Under total ischemia the heart ceased pulsing and neither phosphocreatine nor ATP resonances were detectable but the inorganic phosphate resonance increased. With reperfusion the heart started to beat again and the phosphocreatine and ATP resonances rose above the ischemic levels while the P_1 resonance decreased. The total intensity of ^{31}P resonance decreased with time, presumably due to leakage of P_1 from the heart into the perfusate during ischemia. Tissue pH was monitored by the P_i chemical shift. The intracellular pH decreased by 1.2 pH units during partial ischemia and by an additional 0.5 pH units during total ischemia. Studies on the effects of pacing the heart as well as the time course of development of the effects of ischemia will also be shown. (Supported by USPHS and American Heart Association).

42. NMR CYTOLOGY

D. VUCELIC, N. JURANIC, S. MACURA, S. MACURA and B. NESKOVIC*

Department of Physical Chemistry, University of Belgrade, and Institute of Chemistry, Technology and Metallurgy, Belgrade, Yugoslavia
*Laboratory of Experimental Oncology, University of Belgrade, Belgrade, Yugoslavia

The L-strain 929, HeLa, transformed and normal embryonic human cells, Yoshida sarcoma tumors and C3H mice cells in a 0.8 per cent NaCl solution in heavy water have been investigated by pulse proton NMR. It has been shown that T_1 and T_2 of protons in different cells differ fairly from cell to cell and very much from those of free water. There is a distinct difference in T_1 (also in T_2 but smaller) between different human cells (HeLa, embryonic normal cells and embryonic transformed ones). At the same time, there is a significant difference, between cells of the same kind but of different origin (embryonic normal human cells and embryonic normal cells of the mouse). The interesting observation enables identification of the kind of cells on the basis of their relaxation times (NMR cytology). For purely instrumental reasons, it is somewhat easier to differentiate longitudinal times (T_1), than transverse ones (T_2) and to use these data for cytological identification. It should be taken into account that the signals are derived mostly from water protons indicating that at least one part of water molecules must be strongly oriented. Consequently, ordering of water is characteristic for every kind of cell and can be used as a cellular self-label.

43. TISSUE WATER T1 STUDIES AND HUMAN WHOLE-BODY NMR IMAGING AT ABERDEEN

J.M.S. HUTCHISON, R. SUTHERLAND, J.R. MALLARD and M.A. FOSTER

Department of Medical Physics, University of Aberdeen, Forester Hill, Aberdeen, AB9 2ZD

Water is the most frequently encountered molecule in biological material but the NMR relaxation times arising from its two hydrogen atoms are varied by conditions within the tissue which affect the structuring of the water. Various suggestions have been made for using these variations in the clinical context. In general it has been found that T_1 increases in biopsy samples of cancerous tissue, indicating an increase in free water in such tissue. Suggestions have been made that this would be of value in cancer diagnosis, but other workers have refuted this because overlap exists between values from a population of normal biopsies and a population of malignant biopsies. In the few cases, however, where information has been obtained from unaffected and malignant tissues in the same individual, this increase in T_1 is often quite marked. Because NMR imaging (zeugmatography) presents information from the same individual in one image so that he acts as his own biological control, it may improve the discrimination and demonstrate a real clinical usefulness for these T_1 differences.

Maps of normal values for T_1 in tissues are being prepared, both from published work and from our own work, to assist in the understanding of whole-body NMR images. It is hoped that, as indicated in the literature, other non-malignant changes in tissue will be observed readily by the whole-body technique, e.g. oedema in lung and brain are known to increase T_1, various kidney diseases affect water balance, etc.

The proposed whole-body NMR imaging apparatus being built at Aberdeen works on the principle of selective excitation. Only the spins lying on or close to a chosen horizontal line within the body are excited, using a combination of radio-frequency energy and pulsed magnetic field gradients in directions perpendicular to the line. Immediately following this excitation, the field gradient is switched to lie along this line, and the nuclear induction signal collected and processed. The Fourier transform of the induction signal gives one complete line of the image. Further image lines are obtained by moving the chosen line, either electronically or by moving the patient relative to the apparatus. Information relating to T_1 is obtained by applying a spin-inverting R.F. pulse some time prior to the selective-excitation sequence.

The apparatus operates at a proton resonance frequency below 2 MHz. The resolution of the machine, as far as small changes in T_1 are concerned, is intimately linked with the spatial resolution, expressed as a minimum volume element. We predict a resolution of 10 per cent in T_1 for a volume element of 10 cm^3 but this can be improved by repeating a line a number of times and signal-averaging. We are aiming at a minimum line cross-section of 1 cm^2; resolution along the line is potentially much better at about 3 mm, but at the expense of T_1 accuracy. This resolution limit depends on the overall size of the body under investigation. To investigate the head, we can use a much smaller R.F. coil and hence improve either volume or T_1 resolution by a factor of nearly four.

44. MOLECULAR EVENTS DURING THE INTERACTION OF
ENVELOPED RIBOVIRUSES WITH CELLS. A 270 MHz [1]H
NUCLEAR MAGNETIC RESONANCE STUDY

C. NICOLAU, K. HILDENBRAND, A. REIMANN

*Institut für Strahlenchemie im Max-Planck Institut
für Kohlenforschung, Mulheim/Ruhr, W. Germany*

The ^{1}H NMR spectra of cultured chick embryo fibroblasts
infected with the Newcastle Disease virus, trypsinized
influenza virus and egg-grown influenza virus show considerable broadening of the cell membrane lipid hydrocarbon
chain resonances. These changes are indicative of the
stiffening of the cell membrane lipids. This seems to be
achieved by means of a significant cholesterol transfer
from the virus envelope to the cell plasma membranes, as
radioactivity studies with ^{14}C-cholesterol indicate.

Non trypsinized influenza virus, non infective for the
chick embryo fibroblasts, failed to produce any significant changes in the spectra of the cell membrane lipids.

The Rous Sarcoma Virus, Prague B strain, used with two
lines of chick embryo fibroblasts, one susceptible to the
virus (C/E) and one not susceptible (C/B) showed, after
about 3 hours considerable proteolytic activity, with the
susceptible cells. This activity was evidenced by the
appearance of resolved amino acid spectra; the intensity
of this spectra increased over a period of 8 hours until
the lipid bilayer spectrum of the cell membranes disappeared — probably due to extensive broadening.

Though some proteolytic activity was detected with the
system CEF C/B + RSV too, the lipid bilayer spectrum of
the cell membranes remained unaltered over a period of 10
hours.

These observations permitted some conclusions concerning
the molecular mechanism of penetration of some enveloped
riboviruses into cells.

45. THE HYDRATION PROPERTIES OF SKIN AND ITS COMPONENTS

J.L. MARCHANT, W. DERBYSHIRE and J. CLIFFORD*

*Department of Physics, University of Nottingham,
Nottingham, NG7 2RD, U.K.*
*Unilever Research Laboratories, Port Sunlight, Merseyside, U.K.

The mechanical properties of skin are determined by the
hydration of the stratum corneum outer layer. This in turn
is influenced by three factors, the supply of water from
the deeper layers, the rate of evaporation from the surface, and the water retentive properties of the stratum
corneum itself. This latter quantity shows variability due
to factors such as skin type, state of health etc. The
hydration level can be decreased by the extraction of a

water soluble component of the skin. Measurements of the hydration properties of this water soluble components, and particularly of one component, pyrrolidone carboxylic acid are described. The proton relaxation rates of the water and pyrrolidone carboxylic acid have been recorded as functions of temperature, 213 to 323 K, pH value 0.85 to 10.6 and water content. The hydration behaviour is compared with that of the polypeptide poly L glutamic acid, and with the stratum corneum with and without its water soluble components.

46. TRITIUM NMR SPECTROSCOPY AND ITS APPLICATIONS
IN BIOCHEMISTRY AND BIOLOGY

L.J. ALTMAN and P.C. LAUTERBUR

*Department of Cehmistry, State University of New York
at Stony Brook, Stony Brook, New York, 11794, U.S.A.*

Tritium (T, or ^{3}H) has spin $\frac{1}{2}$, a larger magnetic moment than the proton, negligible natural abundance, and decays by 18.5 keV β-emission with a half-life of 12.26 years. The NMR sensitivity is better than that for protons; there are no background signals to interfere with the detection of even broad or complicated spectra from labelled compounds or their reaction products; and heteronuclear proton decoupling may be used to simplify peaks, increasing the effective resolution and permitting, for example, the study of a conformationally-significant ^{3}H-^{3}H couplings in multiply-tritiated molecules. The extensive use of tritium in biochemical and biological NMR investigations is impeded only by the consequences of its radioactivity, such as radiation hazards, the radiation-induced decomposition of tritiated reagents and experimental solutions, and radiation poisoning of living systems. All of these problems are believed to be amenable enough to solution to permit the extensive and effective use of tritium NMR in studies of complex biological systems, especially those for which low concentrations or interference with the signals of the label by naturally-present material preclude the practical use of ^{1}H, ^{2}H, ^{13}C, ^{15}N or ^{31}P resonances. A number of examples of applications of tritium NMR to the study and analysis of biochemically-interesting compounds will be presented, including quantitative analyses of label positions and distribution patterns, and NOE, T_1 and isotope effect studies.

References

1. Bloxsidge, J., Elvidge, J.A., Jones, J.R. and Evans, E.A. (1971). *Org. Magnetic Res.* **3**, 127.
2. Al-Rawi, J.M.A., Elvidge, J.A., Jaiswal, D.K., Jones, J.R. and Thomas, R. (1974). *J. Chem. Soc. Chem. Comm.* **220**

3. Al-Rawi, J.M.A., Elvidge, J.A., Thomas, R. and Wright, B.J. (1974). *J. Chem. Soc. Chem. Comm.* **1031**.
4. Al-Rawi, J.M.A., Bloxsidge, J.P., O'Brien, C., Caddy, D.E., Elvidge, J.A., Jones, J.R. and Evans, E.A. (1974). *J. Chem. Soc. Perkin II* **1635**.
5. Al-Rawi, J.M.A., Elvidge, J.A., Jones, J.R. and Evans, E.A. (1975). *J. Chem. Soc. Perkin II* **449**.
6. Cary, L.W. (1975). *Rev. Sci. Instr.* **46**, 1422.
7. Al-Rawi, J.M.A., Elvidge, J.A., Jones, J.R., Chambers, V.M.A. and Evans, E.A. (1976). *J. Label. Compounds Radiopharmaceuticals* **12**, 265.
8. Al-Rawi, J.M.A., Bloxsidge, J.P., Elvidge, J.A., Jones, J.R., Chambers, V.M.A. and Evans, E.A. (1976). *J. Label. Compounds Radiopharmaceuticals* **12**, 293.
9. Al-Rawi, J.M.A., Bloxsidge, J.P., Elvidge, J.A., Jones, J.R., Chambers, V.E.M., Chambers, V.M.A. and Evans, E.A. (1976). *Steroids* **28**, 359.
10. Altman, L.J. and Silberman, N. *Anal. Biochem.* (in press).
11. Altman, L.J. and Silberman, N. *Steroids* (in press).
12. Randall, M.H., Altman, L.J. and Lefkowitz, R.J. *J. Med. Chem.* (in press).

47. PROTON MAGNETIC RESONANCE STUDY OF CYTOCHROME C_7

J.J.G. MOURA,* A.V. XAVIER,* G.R. MOORE[†] and R.J.P. WILLIAMS[†]

Centro de Química Estrutual da Universidade de Lisboa, I.S.T., Lisboa 1, Portugal
†*Inorganic Chemistry Laboratory, University of Oxford South Parks Road, Oxford, U.K.*

Desulfuromonas cytochrome C_7 consists of a 68 amino acid polypeptide chain and three haem groups. The protein is related to the tetrahaem cytochromes C_3 from *Desulfovibrio* species. A structure for cytochrome C_3 based on our recent EPR and NMR studies and upon sequence comparisons is proposed.

Cytochrome C_7 contains three -C-X-X-C-H- sequences which act as haem binding regions and three other His residues. None of the residues of the His residues is observable in the 5 to 11 ppm region of the 270 MHz proton NMR spectra of oxidized and reduced cytochrome C_7. These data, coupled with the physical properties and visible spectra of cytochrome C_7 indicate that each iron is bound by two axial His ligands, as with cytochrome C_3.

Further study of the structure of cytochrome C_7 requires assignments for the amino acid resonances in all oxidation states. Cross-assignments for the amino acid resonances of Tyr 6 and some haem meso and amino acid methyl resonances have been obtained from a study of the effect of electron-exchange, between oxidised and reduced cytochrome C_7, upon the NMR spectrum of cytochrome C_7.

Even on the basis of these limited studies it is possible to put forward a tentative fold of the protein chain around the three haem groups. Together with the study on the cytochrome C_3 series it will be possible to analyse internal electron transfer rates in proteins.

AUTHOR INDEX

SUBJECT INDEX

adrenal glands,
  ^{1}H NMR, 278
  ^{31}P NMR, 254
adrenal medulla,
  ^{1}H NMR, 279
  ^{31}P NMR, 253, 279
adrenaline, 253
  interaction with ATP, 255
anaesthetics, 354
antibody, 125
  antigen recognition, 125
  crystal structure, 126, 153
  hypervariable residues, 128
  immunoglobulin fold, 126
  myeloma proteins, 126
  structure prediction of com-
    bining sites, 128
antibody combining site, 125 ff.
  affinity labelling of, 148
  antibody specificity and
    diversity, 151
  assignment of histidine resi-
    dues, 134
  combining site dimensions, 129
  cross saturation, 147
  dependence of hapten orienta-
    tion on binding affinity,
    147
  multispecificity of, 149
  nuclear Overhauser effect, 144
  ^{31}P NMR, use of, 137
  ring currents, use of, 139 ff.
  spin label difference spectro-
    scopy, 134
  the Dnp binding site, 139, 355
aromatic rings, 35
  flipping, 4
  temperature dependence, 42
  see also ring currents
assignment of NMR spectra
  antibody histidine residues,
    134
  aromatic region of the tropo-
    myosin proton spectrum, 163
    ff.

cytochrome c, 341
  first stage, 34–5
  fluorotyrosine gene 5 protein
    ^{19}F spectrum, 175
  general principles, 65 ff.
  lysozyme tyrosine resonances,
    69
  second stage, 35, 37–8
    chemical modification, 38
    X-ray structure, 38

bile salts, 353
brain slices, 385
bufo bufo — muscle studied by
  ^{31}P NMR, 295

carbonic anhydrase, 41, 46, 257
cells,
  normal, 241
  transformed, 241–4
  tumour, 242
chemical exchange,
  broadening of tyrosine, 41
  lysozyme tryptophan N(1)H pro-
    tons, 71
chemical modifications,
  affinity labelling of an anti-
    body combining site, 148
  conformational change on modi-
    fication of lysozyme trypto-
    phan, 87, 88, 108
  gene 5 protein, tyrosines and
    lysine, 174
  histones, 343
chemical shift anisotropy, 43,
  235, 239
chromaffin granules,
  ^{1}H NMR, 280
  ^{31}P NMR, 253–8
chromogranin A, 281
chymotrypsin, 1, 2
^{35}Cl$^-$ NMR, 258
conalbumin, 360
conformational changes,
  absence of on antibody-hapten